LESSON TESTS AND ANSWER KEYS

FRIENDLY

BIOLOGY

Joey Hajda DVM MEd   Lisa B. Hajda

Published by Hideaway Ventures, 79372 Road 443, Broken Bow, NE 68822

For information regarding this publication please contact Joey or Lisa Hajda at the above address or visit our website: www.friendlybiology.com.

Copyright 2016 Hideaway Ventures, Joey or Lisa Hajda

TABLE OF CONTENTS

NAME_____DATE_____
FRIENDLY BIOLOGY
LESSON 1 TEST

Read each question below carefully. Choose the one best answer. Write the letter in the space provided. There may be some responses which refer to more than one response being correct. Read carefully. An answer with a misspelled response is a poor choice.

_____1. The term biology is derived from two terms: bios- and -ology. The portion bios- refers to:

A. Cellular components
B. Life
C. Insects
D. Blood

_____2. A cat is chasing a mouse across the floor of the house. This is evidence that:

A. The cat and the mouse are living things.
B. The cat is a living thing but the mouse is not.
C. This is not evidence that the cat or the mouse is a living thing.
D. The floor is a living thing.

_____3. Which item below is evidence that plants do indeed move?

A. A plant drops its leaves in the fall.
B. A plant repositions its leaves during the day to get maximum exposure to the sun.
C. A plant's flower heads move when air currents move past.
D. A plant's roots are capable of lifting a section of sidewalk.

_____4. The process whereby a living creature is capable to creating a new living creature is known as:

A. Duplication
B. Carbon copying
C. Reproduction
D. Replication

_____5. The need for your pet to be fed on a daily basis is an example of:

A. Living things requiring an energy source
B. Living things being given something on a daily basis.
C. Living things responding to their environment.
D. Living things reproducing.

_____6. Plants have an advantage over animals, including humans, in that plants:

A. Are able to consume non-living sources of energy.
B. Are able to create their own source of food.
C. Are able to reproduce with plants other than their own kind.
D. Cannot move and therefore cannot be chased by cats.

_____7. Which of the following is an example of the feature that living things develop and grow?

A. Puppies enlarge in size and become dogs.
B. Feathers replace down on chicks.
C. Young heifers gain the capability of having calves.
D. A fawn loses its spotted color pattern on its fur.
E. All are examples of living things developing and growing.

_____8. When a board is lifted from the grass, several visible earthworms suddenly begin to burrow into the moist soil. This is an example of living things:

A. Reproducing
B. Growing and developing
C. Responding to their environment.
D. Requiring a source of energy.

_____9. A rock sinks to the bottom of the lake after being tossed in by a fishermen. Which of the following statements could be considered true regarding living and non-living things?

A. The water splashed when the rock entered the surface and can therefore be considered a living thing.
B. The rock moved to the bottom of the lake and this fact is evidence that the rock is a living thing.
C. The fisherman propelled the rock through the air which is evidence that he is a living thing.
D. The fish avoided the falling rock by swimming. This is evidence the fish are living things.
E. Both choices c and d are good choices.

_____10. The leaves on a maple tree emerge from buds in April in Midwestern states. This is a good example that living things:

A. Reproduce
B. Develop and grow
C. Move
D. Are present in all states of the US.

NAME_____DATE_____
FRIENDLY BIOLOGY
LESSON 2 TEST

Read each question below carefully. Choose the one best answer. Write the letter in the space provided. There may be some responses which refer to more than one response being correct. Read carefully. An answer with a misspelled response is a poor choice.

_____1. All things, whether living or not, are composed of tiny bits of matter known as:

A. Atoms
B. Crumbs
C. Propellant particles
D. Energized photons

_____2. The atomic theory says that atoms are composed of yet smaller subatomic particles. The subatomic particles found in the nucleus of the atom are the

A. Electrons and protons
B. Electrons and neutrons
C. Electrons
D. Protons and neutrons

_____3. The number of electrons found in an atom can be found on the periodic table of elements. This value is known as the:

A. Atomic mass number
B. Atomic number
C. Element symbol
D. Electron mass number

_____4. The arrangement of the _____ tells us about the behavior or reactivity of an element's atoms.

A. Protons
B. Neutrons
C. Electrons
D. Croutons

_____5. Which statement below is correct regarding the behavior of atoms of various elements?

A. Elements whose atoms have their outermost layer of protons completely filled are most stable.
B. Elements which have their nucleus filled with electrons are the most stable.
C. Elements which have their outermost layer of electrons partially filled are the most stable.
D. Elements which have their outermost layer of electrons completely filled are the most stable elements.

_____6. Elements which are unstable are able to gain stability by forming _____ with other unstable elements.

A. Business relationships.
B. Partnerships
C. Bonds
D. Stocks and mutual funds

_____7. When atoms transfer electrons from one to another in order to form greater stability, a(n) _____ bond is said have formed.

A. Ionic
B. Ironic
C. Covalent
D. Life-long

_____8. When atoms share electrons, rather than transfer electrons, to achieve greater stability, _____ bonds are said to have formed.

A. Ionic
B. Covalent
C. Neutronic
D. Savings

_____9. Oxygen atoms share two electrons and therefore the bond formed is called a:

A. Single ionic bond
B. Double ionic bond
C. Triple covalent bond
D. Double covalent bond

_____10. Four elements which are vitally important to living things are:
A. Carbon, hydrogen, helium and neon
B. Calcium, hydrogen, oxygen and nickel
C. Carbon, hydrogen, oxygen and nitrogen
D. Carbon, hydrogen, oxigen and nitrogin

_____11. Carbon atoms desire to form ___ bonds while oxygen atoms desire to form _____ bonds to gain greater stability.
A. Two, four
B. Six, two
C. Four, six
D. Four, two

_____12. Which of the following is not considered a feature common to all living things (that was presented in Lesson 1):
A. Living things grow and develop.
B. Living things require a source of food or energy.
C. Living things move.
D. All living things require oxygen for survival.
E. Living things reproduce.
F. Living things respond to their environment.

NAME_____DATE_____

FRIENDLY BIOLOGY

LESSON 3 TEST

Read each question below carefully. Choose the one best answer. Write the letter in the space provided. There may be some responses which refer to more than one response being correct. Read carefully. An answer with a misspelled response is a poor choice.

_____1. The common name for carbohydrates is:

A. Aldehydes
B. Sugars
C. Carbonated beverages
D. Fats

_____2. Carbohydrates consist of three elements which are:

A. Carbon, hydrogin and oxygen
B. Carbon , hydrogen and oxygin
C. Carbon, hydrogen and oxygen
D. Calcium, hydrogen and oxygen

_____3. The generic formula for a carbohydrate is

A. $C_n(H_2O)_n$
B. $C_6(H_2O)_n$
C. $C_n(H_{22}O)_n$
D. $C_6(H_{12}O)_6$

_____4. The most important sugar to living things is _____

A. Sucrose
B. Glucose
C. Maltose
D. Lactose

_____5. When one considers the chemical formula for glucose, n = _____

A. 12
B. 2
C. 8
D. 6

_____6. Carbohydrates generally have a _____ structure.

A. Ring
B. Triangular
C. Linear
D. Pyramidal

_____7. The scientific name for blood sugar is:

A. Lactose
B. Cellulose
C. Glucose
D. Fructose

_____8. Glucose and fructose both share n = 6, however, they do not have the same struc-
ture. They are known as _____.

A. Icingles
B. Isomers
C. Mirror images
D. Diametric structures

_____9. Which food below would be unlikely to contain fructose?

A. Apples
B. Beef
C. Oranges
D. Plums

_____10. Choose from the carbohydrates listed below, the one naturally found in dairy prod-
ucts.

A. Sucrose
B. Maltose
C. Cellulose
D. Lactose

_____11. Regarding the carbohydrate, sucrose, choose the correct statement.

A. Sucrose is commonly known as milk sugar.
B. Sucrose is a monosaccharide.
C. Sucrose is commonly known as table sugar and is a disaccharide.
D. Sucrose is the sugar found in fruits.

_____12. Lactose is a disaccharide. The two carbohydrates which made up lactose are:

A. Glucose and lactose
B. Glucose and galactose
C. Fructose and maltose
D. Amylose and pentose

_____13. The carbohydrate which functions like "gasoline" for our bodies is

A. Lactose
B. Sucrose
C. Glucose
D. Enzyme

_____14. Enzymes allow our bodies to take the foods we eat and make them useful. The two main jobs that enzymes do are:

A. Make reactions happen more slowly and to cut compounds apart.
B. Make reactions happen more rapidly with themselves not being changed and to cut compounds apart.
C. Adjust levels of various compounds in the body making their concentrations correct.
D. Quicken reactions and allow for rapid growth and development.

_____15. Which description below best describes a person who is diagnosed as being lactose intolerant?

A. The person lacks sufficient lactase to adequately digest dairy products. This results in gas production from bacteria present in the digestive tract which consume the excess lactose.
B. The person produces an excess of lactase which results in gas production and discomfort. Taking lactase inhibitors can reduce symptoms.
C. The person lacks sufficient lactose in their diet and must supplement with additional dairy products.
D. The person must not eat dairy products due to an increase in sucrase enzyme interactions.

_____16. The carbohydrate found in grains is known as:

A. Heptose
B. Sucrose
C. Maltose
D. Lactose

_____17. Plants have the ability to assemble large amounts of glucose into storage for later use. This storage form of glucose is known as

A. Amylose, commonly known as starch.
B. Glucase, commonly known as blood sugar.
C. Cellulose, commonly known as starch.
D. Maltose, commonly known as grain sugar.

_____18. The process of creating glucose in plants through the use of light energy from sunlight, is known as:

A. Amylosynthesis
B. Photosinthesis
C. Respiration
D. Photosynthesis

_____19. The enzyme responsible for breaking down amylose into constituent glucose molecules is known as:
A. Glucase
B. Fructase
C. Amylase
D. Lactase

_____20. Plants are also able to link glucose molecules into long, strong chains to give strength and support to the plant. This carbohydrate is known as _____ and the enzyme which works to break it down back into glucose molecules for use is _____.

A. Cellulose, cellulase
B. Lactose, lactase
C. Amylose, amylase
D. Fructose, fructase

_____21. Animals which have microorganisms in their digestive tracts capable of breaking down cellulose are known as _____. The relationship that exists between these animals and the microorganisms is known as _____.

A. Predators, parasitism
B. Ruminants, mutual symbiotic relationship
C. Herbivores, mutual parasitic relationship
D. Carnivores, symbiosis

_____22. Fleas, ticks, mosquitos are examples of living things which live at the expense of another living thing. The relationship which exists between these living things and the host species from which it gains benefit is known as:

A. A mutual symbiotic relationship.
B. Commensalism
C. Parasitism
D. A bad relationship.

NAME_____DATE_____
FRIENDLY BIOLOGY
LESSON 4 TEST

Read each question below carefully. Choose the one best answer. Write the letter in the space provided. There may be some responses which refer to more than one response being correct. Read carefully. An answer with a misspelled response is a poor choice.

_____1. Which statement below about lipids is the false statement?

A. Lipids are commonly known as fats.
B. Lipids are small, simple molecules.
C. Lipids consist of two parts: glycerol portion and fatty acid portion.
D. Lipids are composed of carbon, hydrogen and oxygen atoms.

_____2. The glycerol portion of a lipid contains _____ carbon atoms.

A. One
B. Two
C. Three
D. Several

_____3. A lipid with one fatty acid chain attached to the glycerol portion is known as a:

A. Monosacharride
B. Diglyceride
C. Monoglyceride
D. Tryglyceride

_____4. A fatty acid chain which is straight and has all carbon atoms "filled' with hydrogen atoms is known as a(n):

A. Saturated fatty acid
B. Unsaturated fatty acid
C. Polysacharride
D. Pre-filled fatty acid

_____5. In unsaturated fatty acids, one will find a _____.

A. Double-bonded set of hydrogen atoms
B. Double-bonded set of carbon atoms
C. Pair of oxygen atoms which are covalently bonded together.
D. Long chain of carbon atoms all with single bonds present.

_____6. Saturated fats would most likely be found in which food product below?

A. Bacon
B. Corn oil
C. Peanut butter (with no added animal fats)
D. Palm oil

_____7. On a table in your house at room temperature you have been given two fats. Fat A is solid and Fat B is liquid. Choose the statement below that best describes these fats.

A. Fat A is most likely a saturated fat and is from animal origin.
B. Fat A is most likely an unsaturated fat and is from animal origin.
C. Fat B is most likely a saturated fat and is from animal origin.
D. Fat B is most likely a saturated fat and is from plant origin.

_____8. Which statement below best describes how lipids react with water?

A. Lipids mix readily with water.
B. Lipids are insoluble in water due to the way the hydrogen and oxygen atoms are arranged.
C. Lipids are soluble in water due to the way the hydrogen and oxygen atoms are arranged.
D. Lipids mix readily with water once they reach body temperature.

_____9. Which term below best describes a lipid?

A. Hydrophilic
B. Hydrophobic
C. Arachniphilic
D. Arachniphobic

_____10. Lipids can be considered like large stockpiles of _____.

A. Glucose molecules
B. Lactose molecules
C. Starch molecules
D. Greasy frenchfries.

_____11. The enzymes which work to lyse lipids into smaller, useful components are called:

A. Fat knives
B. Fat scissors
C. Amylase enzymes
D. Lipases

_____12. One cup of a lipid has about _____ times the amount of energy found in the same amount of glucose.

A. 10
B. 9
C. 100
D. 900

_____13. The process by which the lipid portion of milk is made to remain equally distributed throughout the watery portion of milk is known as _____.

A. Filtration
B. Pasteurization
C. Cremation
D. Homogenization

FRIENDLY BIOLOGY

LESSON 5 TEST

Read each question below carefully. Choose the one best answer. Write the letter in the space provided. There may be some responses which refer to more than one response being correct. Read carefully. An answer with a misspelled response is a poor choice.

_____1. Proteins, like carbohydrates and lipids, contain carbon, hydrogen and oxygen atoms. In addition to these three, proteins also contain:

A. Nitrogin
B. Nitrogen
C. Nickel
D. Neptunium

_____2. Proteins are built of smaller subunits known as

A. Archaic acids
B. Amino acids
C. Arachidonic acids
D. Weak acids

_____3. Amino acids consist of two portions: the _____ portion and the change-able _____.

A. Common, r-group
B. Common, p-group
C. Same, x-group
D. Same, carboxyl

_____4. It's the _____ of the amino acid that determines the identity of the amino acid.

A. Carboxyl group
B. Hydroxyl group
C. R-group
D. Amino group

_____5. Of all the amino acids, there are eight of special importance to humans and must be taken in our daily diet. These amino acids are referred to as the:

A. Special amino acids
B. Essential amino acids
C. Rare earth amino acids
D. R-group amino acids

_____6. Proteins are linked together through the use of _____, known as _____.

A. Enzymes, peptidases
B. Enzymes, lipases
C. Carbohydrates, oxalases
D. Lipids, lipases

_____7. The process of linking amino acids together in specific sequences to build proteins is known as _____ due to the release of a molecule of _____.

A. Respiration, carbon
B. Dehydration emphasis, carbon
C. Dehydration synthesis, water
D. Respiration synthesis, water

_____8. Many amino acids linked together form what is referred to as a _____.

A. Carbohydrate
B. Dipeptide
C. Peptide bond
D. Polypeptide or protein

_____9. The enzymes which link amino acids are known as _____.

A. Peptidases
B. Lipases
C. Lactases
D. Amino acids

_____10. Choose the statement below which is correct regarding the breaking apart of proteins consumed in our diet.

A. Polypeptides are broken apart by enzymes known as peptidases.   In order to complete the process, water is required. The process is known as hydrolysis.
B. Polypeptides are broken apart by enzymes known as lipases. A molecule of water is required and the process is known as dehydration synthesis.
C. Proteins are broken apart by enzymes known as pastorases. Because water required, the process is known as dry hydration degradation.
D. Proteins are broken down by carbohydrates in solutions of water and this process is known as respiration.

_____11. Which statement below best describes the linking together of amino acids?

A. When two amino acids are linking together a polypeptide forms.
B. Many amino acids which have been linked together are called a polypeptide.
C. The bond between linking amino acids is known as an ionic bond.
D. A tripeptide consists of two linked amino acids.

_____12. Which statement below, in general, states the primary role of proteins in living things?

A. The primary role of proteins in living things is to serve as movers of fluids throughout cells.
B. The primary role of proteins in living things is to serve as a means to provide a source of energy or fuel for the body.
C. The primary role of proteins in living things is to allow living things to make copies of genetic materials.
D. The primary role of proteins in living things is provide a source of building materials for structure.

_____13. Which statement below is not a true statement regarding proteins in living things?

A. Albumen is a protein found in egg whites and also in the blood.
B. Immunoglobulins are proteins which are vital to the function of the immune system.
C. The first milk found in mammals is known as colostrum and contains proteins important to the survival of the baby mammal.
D. Proteins, like carbohydrates, are small molecules consisting of only a few atoms of each element (carbon, hydrogen, oxygen and nitro-
gen.)

FRIENDLY BIOLOGY

LESSON 6 TEST

Read each question below carefully. Choose the one best answer. Write the letter in the space provided. There may be some responses which refer to more than one response being correct. Read carefully. An answer with a misspelled response is a poor choice.

_____1. In the term pH, the symbol H indicates

A. Helium ion
B. Hydrogen ion
C. Hydrogen crouton
D. Hafnium ion

_____2. Acids have hydrogen ions which

A. Desire to leave the acid to gain better taste.
B. Desire to leave the acid in order to gain greater stability.
C. Desire to not leave the acid to gain greater stability.
D. Desire to not leave the base to gain greater stability.

_____3. Stomach acid is scientifically known as _____.

A. Acetic acid
B. Vinegar
C. Hydrochloric acid
D. Battery acid

_____4. The pH scale extends from _____ with bases having the _____ numbers.

A. 10-25, lower
B. 0-20, lower
C. 0-10, higher
D. 0-14, higher

_____5. Bases have the desire to

A. Lose hydrogen ions
B. Accept hydrogen ions
C. Not gain nor lose hydrogen ions
D. Accept helium ions

_____6. Bases have _____ numbers on the pH scale and taste _____.

A. High, sour
B. High, bitter
C. Low, sour
D. Low, bitter

_____7. A value of ___ is considered neutral on the pH scale while the pH of human blood is
_____.

A. 6.5   7.0
B. 7.0   6.5
C. 7.4   7.8
D. 7.0   7.4

_____8. All portions of the human body have the pH of 7.4

A. True
B. False

_____9. The pH scale is a logarithmic scale. This means that

A. One step up on the pH scale is 10 times the previous value.
B. One step up on the pH scale is 100 times the previous value.
C. One step up on the pH scale is double the previous value.
D. One step up on the pH scale is one-tenth the previous value.

_____10. Suppose you had a substance which turned red litmus paper a blue color. Which statement below agrees with this finding?

A. The substance is most likely an acid.
B. This substance is most likely a base.
C. This substance would most likely have sour taste.
D. This substance would most likely have a pH value of less than 7.0.

_____11. Understanding pH is important because

A. Enzymes in living things function best at specific pH levels.
B. All enzymes function best at acidic pH levels.
C. Proteins function best at very low pH levels.
D. The immune system is dependent upon the body maintaining highly basic conditions.

NAME_____DATE_____
FRIENDLY BIOLOGY

LESSON 7 TEST

Read each question below carefully. Choose the one best answer. Write the letter in the space provided. There may be some responses which refer to more than one response being correct. Read carefully. An answer with a misspelled response is a poor choice.

_____1. The study of cells is known as

A. Histology
B. Chytology
C. Cytology
D. Geology

_____2. The "skin" of cells is known as the _____ and it is composed of _____.

A. Nucleus, a phospholipid bilayer (two layers)
B. Mitochondrion, chemolipid bilayer
C. Plasma or cell membrane, phospholipid bilayer
D. Nuclear membrane, phospholipid monolayer (single layer)

_____3. The lipid portion of the cell membrane is considered to be

A. Hydrophilic
B. Hydrophobic
C. Thermophilic
D. Thermophobic

_____4. The cell membrane functions like a _____ with _____.

A. Storage closet with doors.
B. Brain with gates.
C. Fence with gates.
D. Wall with pockets.

_____5. Which statement below best describes the cell wall in living things?

A. Cell walls, found in plants and animals, function like the skeleton of the cell.
B. Cell walls, found in plants, consist of lactose and fructose.
C. Cell walls, found only in animals, consists primarily of amylose.
D. Cell walls, found in plants, consist of varying degrees of cellulose.

_____6. Which organelle listed below functions like the brain or control center for the cell?

A. Nucleus
B. Nuclear envelope
C. Mitochondrion
D. Cell membrane

_____7. Bacteria have no nuclear membranes. They are considered as being

A. Eukaryotes
B. Prokaryotes
C. Unikaryotes
D. Idon'tknowyotes

_____8. The information required for a cell to maintain itself as well as perform its job is found in the cell's

A. Cell membrane
B. Chromosomes
C. Nuclear membrane
D. Mitochondrion

_____9. Genes within cells are

A. Like recipes within chapters (chromosomes) of the cell.
B. Found in the mitochondria of cells and function to limit energy consumption.
C. Are only found in eukaryotic cells.
D. Found only in odd-numbered quantities.

_____10. Humans have _____ pairs of chromosomes in almost all cells.

A. 46
B. 23
C. 21
D. 42

_____11. Living things are incapable of reproducing with other living things other than their own kind primarily because

A. The number and content of each chromosome is specific for each living creature.
B. The size of each chromosome is specific for each living creature.
C. The color of each chromosome is specific for each living creature.
D. Chromosomes are only found in plants and not other living creatures.

_____12. The organelle of cells which is responsible for the conversion of glucose to energy containing compounds is the

A. Nucleus
B. Cell membrane
C. Mitochondrion
D. Mitachondrium

More on next page.

_____13. The process whereby glucose molecules are converted into energy-containing compounds known as ATP's is

A. Hydrolysis
B. Respiration
C. Dehydration synthesis
D. Breathing

_____14. The degree to which the mitochondrion is capable of converting glucose to ATPs is dependent upon the availability of

A. Hydrogen
B. Water
C. Oxygen
D. Lactose

NAME_____DATE_____
FRIENDLY BIOLOGY
LESSON 8 TEST

Read each question below carefully. Choose the one best answer. Write the letter in the space provided. There may be some responses which refer to more than one response being correct. Read carefully. An answer with a misspelled response is a poor choice.

_____1. The main job of the golgi body in cells is to

A. Storage of products created by the cell
B. To control all activities of the cell
C. The package and export products made by the cell
D. Produce energy-containing compounds from fuel sources like glucose.

_____2. In cells, the vacuole may have two possible functions. Those functions are to

A. Produce ATPs and store them.
B. Move substances in and out of cells and control cell activities.
C. Package and "ship" products made by the cell.
D. Store products made by the cell or assist with locomotion of the cell.

_____3. In which cell organelle might one expect to find enzymes?

A. Lysosome
B. Lisosome
C. Golgi body
D. Vacuole

_____4. The main job of the endoplasmic reticulum is

A. Transportation
B. Storage of products made by the cell.
C. Production of ATPs
D. Control of all cell activities

_____5. Smooth and rough endoplasmic reticulum are found in cells. Rough endoplasmic reticulum has _____ on its surface.

A. Lysosomes
B. Ribosomes
C. Mitochrondria
D. Vacuoles

_____6. Linking amino acids into polypeptides is the main function of _____ found on _____.

A. Lysosomes, RER
B. Mitochondria, SER
C. Ribosomes, SER
D. Ribosomes, RER

_____7. Correct folding of proteins is important because

A. Protein function is dependent upon correct folding of proteins.
B. The length of a protein needs to be shortened to increase its ability to control cell activities.
C. The shape of protein is affected by correct folding but folding does not affect protein function.
D. Proteins are not folded, they are allowed to stretch for miles within cells.

_____8. The degrees of protein folding from most simple to most complex are

A. Primary, tertiary, secondary, quaternary
B. Primary, secondary, tertiary, fourth
C. Tertiary, quaternary, secondary, primary
D. Primary, secondary, tertiary, quaternary

_____9. Sarcoplasmic reticulum is found in _____ and functions to maintain adequate levels of the element _____.

A. Heart, sodium
B. Muscles, sodium
C. Stomach, hydrogen
D. Muscles, calcium

_____10. The ability for a cell to contract and do so when dividing, is due to action of cellular _____.

A. Membranes
B. ER
C. Microtubules
D. Nuclei

_____11. Which statement below best describes chloroplasts found in cells?

A. Chloroplasts are found in plant cells and function to convert carbon dioxide and water into glucose.
B. Chloroplasts are found in animal cells and function to convert glucose into ATPs.
C. Chloroplasts are responsible for the conversion of light energy into carbon dioxide and water in order to provide a food source for the plant.
D. Chloroplasts are responsible for the conversion of carbon monoxide into lactose in the process known as photosynthesis.

_____12. A byproduct of photosynthesis is _____ which is vital to the survival of other creatures like human beings.

A. Oxygen
B. Carbon dioxide
C. Sunlight
D. Water

NAME_____DATE_____
FRIENDLY BIOLOGY
LESSON 9 TEST

Read each question below carefully. Choose the one best answer. Write the letter in the space provided. There may be some responses which refer to more than one response being correct. Read carefully. An answer with a misspelled response is a poor choice.

_____1. Cells divide into more cells through a process known as

A. Mytosis
B. Yourtosis
C. Ourtosis
D. Mitosis

_____2. Which statement below best describes the period of interphase within the life cycle of a cell?

A. Interphase is the period of time when a cell is making copies of its chromosomes.
B. Interphase is the period of time when a cell is doing its own assigned job and not doing any cell division.
C. Interphase is the period of time when the chromosomes within a cell are aligning themselves along the "equator" of the nucleus.
D. Interphase is the period of time when the cell membrane begins to pinch inward.

_____3. When one looks at the process of cell division, the cell that does the dividing is referred to as the _____ while the two resulting cells are known as the

_____.

A. Old cell, young cells
B. Parent cell, offspring cells
C. Mother cell, daughter cells
D. Parent cell, daughter cells

_____4. In humans, the signal for a cell to divide may come in the form of growth hormone. Where, in humans does growth hormone arise and where does it have its effects?

A. It arises from the pituitary gland and has its effects all across the body.
B. It arises from the brain and has its effects on bones and muscles only.
C. It arises from the reproductive organs and has its effects upon the pituitary gland.
D. Growth hormone is taken orally in order to have its effects take place in the body.

_____5. Which statement below best describes the events of prophase?

A. In prophase, chromatin condenses back into chromosomes, the chromosomes make copies of themselves in order for each new cell to have their own set and, if present, the nuclear membrane begins to fall apart.
B. In prophase, the chromosomes make copies of themselves, align themselves along the equator of the cell and then begin to move apart.
C. In prophase, the cell is doing its own assigned tasks.
D. In prophase, the cell begins to split into two new cells.

_____6. In this stage of mitosis, the chromosomes align themselves along a plate or equator across the cell and spindle fibers attach themselves. Which stage of mitosis is this?

A. Prophase
B. Telophase
C. Metaphase
D. Anaphase

_____7. Following metaphase, the chromosomes begin migrating to opposite poles of the cell due to the contraction of the spindle fibers. This phase of cell division is known as

A. Mitosis
B. Metaphase
C. Anaphase
D. Telophase

_____8. Which choice below best describes the events of telophase?

A. The chromosomes uncoil themselves once again and the cell membrane begins pinching inward.
B. The chromosomes coil themselves to become visible and make copies of themselves.
C. The chromosomes align themselves along the equator of the cell and being moving to opposite poles of the cell.
D. The chromosomes contact neighboring cells via their i-phones.

_____9. The process whereby the cell membrane pinches inward to eventually create to new cells is known as

A. Mitosis
B. Cytokinesis
C. Citokinesis
D. Cytoplasmic lysing

_____10. If a cell is in interphase and receives the signal to begin mitosis, which set of phases listed below is the correct sequence?

A. Interphase, telophase, anaphase, metaphase, prophase
B. Interphase, prophase, anaphase, metaphase, telophase
C. Interphase, prophase, metaphase, anaphase, telophase
D. Interphase, prophase, matophase, anophase, telophase

FRIENDLY BIOLOGY

LESSON 10 TEST

Read each question below carefully. Choose the one best answer. Write the letter in the space provided. There may be some responses which refer to more than one response being correct. Read carefully. An answer with a misspelled response is a poor choice.

_____1. Segments of chromosomes which carry the specific information for a feature of a living creature are known as _____.

A. Ribosomes
B. Genes
C. Jeans
D. Nuclei

_____2. Genes consist of chains of molecules known as _____.

A. DAN
B. RNA
C. DNA
D. Ribonucleic acid

_____3. DNA is thought to be shaped like a ladder. The rails of the ladder consist of

A. Alternating base pairs.
B. Alternating ribose and phosphate groups.
C. Oxygen and carbohydrate atoms.
D. Alternating Watson and Crick atoms

_____4. The rungs of the DNA ladder consist of

A. Base pairs
B. Ribose and phosphate groups
C. Guanine and ribose groups
D. Adenine and glucose groups

_____5. Adenine and guanine are bases identified as _____.

A. Purines
B. Pyrimidines
C. Strong bases
D. Weak bases

_____6. Thymine and cytosine are bases identified as _____.

A. Purines
B. Pyrimidines
C. Strong bases
D. Weak bases

_____7. The base adenine always bonds with _____.

A. Cytosine
B. Guanine
C. Thymine
D. Adenine

_____8. The base cytosine always bonds with _____.

A. Cytosine
B. Guanine
C. Thymine
D. Adenine

_____9. The bonds found between the base pairs are _____ bonds and are known as _____.

A. Strong, ionic bonds
B. Strong, covalent bonds.
C. Weak, hydrogen bonds.
D. Weak, savings bonds.

_____10. DNA segments are unzipped by

A. Enzimes
B. Ribose sugars
C. Enzymes
D. Covalent bonds

_____11. Which statement best describes how a segment of DNA is replicated?

A. The segment of DNA which is to be copied is unzipped by a lipase enzyme which allows free-floating fatty acids to move in to create the copy.
B. The segment of DNA which is to be copied is unzipped by enzymes which allows free-floating nucleotides to move in, matching the base pairs.
C. The segment of DNA which is to be copied is unzipped by lactase which exposes ribose pairs. Free-floating base pairs move in to match the appropriate base pairs.
D. The segment of DNA to be replicated is folded into a tertiary fold allowing base pairs to be shielded from enzymes which allows copies to be made.

NAME_____DATE_____
FRIENDLY BIOLOGY
LESSON 11 TEST

Read each question below carefully. Choose the one best answer. Write the letter in the space provided. There may be some responses which refer to more than one response being correct. Read carefully. An answer with a misspelled response is a poor choice.

_____1. The very first step in protein synthesis is

A. Creation of base pairs in specific sequences.
B. Unzipping of the desired section of DNA by enzymes.
C. Unzipping of the desired section of double-stranded RNA by enzymes.
D. The floating-in of appropriate base pairs.

_____2. Which statement below regarding RNA is false?

A. RNA is the abbreviation for ribonucleic acid.
B. RNA is single-stranded.
C. RNA has the same bases as DNA.
D. RNA will have the base uracil rather than the base, thymine.

_____3. Where in the cell will one usually find the DNA segment which must be utilized to create proteins?

A. Cytoplasm
B. Mitochondrion
C. Nucleus
D. Golgi body

_____4. When RNA molecules align themselves with exposed DNA bases, the RNA base uracil will bond with _____.

A. Cytosine
B. Guanine
C. Thymine
D. Adenine

_____5. The strand of RNA which forms in the nucleus of cells in an effort to transport this information out to the location of protein synthesis is known as _____.

A. mDNA
B. tRNA
C. MRNA
D. mRNA

_____6. The actual assembly of proteins takes place
A. In the ribosomes on the rough ER.
B. In the vacuoles on the smooth ER.
C. Vesicles of the rough ER.
D. At the hospital in the ER.

_____7. The messenger RNA strand codes for specific amino acids because it can be divided into three-base segments known as _____.

A. Radons
B. Exxons
C. Codons
D. Base trios

_____8. Amino acids which are ready to be assembled into proteins have segments of _____ attached to them which align themselves with segments of mRNA.

A. tRNA
B. mRNA
C. DNA
D. tDNA

_____9. The process in which information is communicated from DNA to mRNA within the nucleus of of a cell is known as _____.

A. Transcription
B. Translation
C. Transegenation
D. Copy facilitation

_____10. The process in which information is communicated from mRNA to the ribosomes is known as _____.

A. Transcription
B. Translation
C. Trenslation
D. Replication facilitation

_____11. Which statement below best describes the results of a genetic mutation?

A. The protein which results from this change may not function correctly.
B. Genetic mutations occur frequently and rarely have any effects upon the living creature.
C. Genetic mutations result in DNA strands which are unique and offer new improvements in a species.
D. Genetic mutations only improve the capabilities of a living thing.

_____12. Radiation from the sun, exposure to radioactive elements or certain chemicals which cause alterations in sequences of DNA or RNA are known as _____.

A. Mutagens
B. Antibiotics
C. Chemical teratogens
D. Growth inhibitors

NAME_____DATE_____
FRIENDLY BIOLOGY
LESSON 12 TEST

Read each question below carefully. Choose the one best answer. Write the letter in the space provided. There may be some responses which refer to more than one response being correct. Read carefully. An answer with a misspelled response is a poor choice.

_____1. The process by which living creatures create more complete living creatures is known as:

A. Mitosis
B. Reproduction
C. Replication
D. Synthesis

_____2. The information necessary to create a new living creature is found in one's chromosomes. In humans there are:

A. 46 pairs of chromosomes
B. 23 pairs of chromosomes, one member of each pair coming from each parent of that individual
C. 48 total chromosomes, one member of each pair coming from each parent of that individual
D. 22 pairs of chromosomes, 20 pairs coming from the mother of the individual and 2 coming from the father of the individual

_____3. Segments of chromosomes which carry information for specific traits or features are known as genes. Variations of a gene, like specific eye color or face shape, are known as

A. Gene markers
B. Alleles
C. Aleles
D. Nuclear codes

_____4. Organisms which have pairs of chromosomes arising from two parents are known as being

A. Coploid
B. Haploid
C. Diploid
D. Triploid

_____5. Diploid cells are indicated by the symbol _____ where haploid cells are indicated by the symbol _____.

A. 2N, 1N
B. 1N, 2N
C. 1A, 2A
D. AA, BB

_____6. Primordial sex cells, whether male or female, are _____ before undergoing meiosis.

A. 1N
B. 2N
C. Haploid
D. In telophase

_____7. Which statement below best describes the process of meiosis?

A. In meiosis, the number of chromosomes found in 2N primordial sex cells is reduced to 1N.
B. In meiosis, the number of chromosomes found in 1N primordial sex cells is doubled to create 2N gametes.
C. In meiosis, the number of chromosomes is reduced by half so that the resulting cells are 2N and no longer 4N.
D. In meiosis, two daughter cells result with exactly the same number of chromosomes as the parent cell.

_____8. In human males meiosis to create sex cells takes place in the _____ whereas in females, meiosis occurs in the _____.

A. Ovaries, testis
B. Testis, kidneys
C. Testis, ovaries
D. Ovaries, kidneys

_____9. In human males, one primordial sex cell results in _____ gametes or sperm cells whereas in females, one primordial sex cell results in _____ .

A. Two, four
B. Four, four ova
C. Four, 1 ovum and three polar bodies
D. Two, four ova and three polar bears

_____10. Which statement below is true regarding the creation of sperm and ova in humans.

A. Males are capable of continually creating more primordial sex cells and therefore more ova throughout their lives.
B. Women are born having all of the primordial sex cells they will have in their lives and are therefore limited in the number of ova they can produce in their lifetime.
C. Men are born having all of the primordial sex cells they will have in their lives and therefore are limited as to the number of sperm cells they can produce in their lifetime.
D. Both men and women are unlimited as to the number of sex cells they can produce in their lifetime.

_____11. The process of creating ova from 2N cells is known as

A. Oogenesis
B. Mitosis
C. Spermatogenesis
D. Photosynthesis

_____12. When the male 1N cell joins the female 1N cell, the process is known as

A. Meiosis
B. Mitosis
C. Spermatogenesis
D. Fertilization

_____13. Sexual reproduction

A. Results in individuals having traits or features unlike either parent involved.
B. Is where two parents, a male and female, are involved in the creation of the new living creature.
C. Results in individuals that are identical to the single parent involved.
D. Only occurs in very low living things like bacteria and protists.

_____14. Bacteria utilize a process of asexual reproduction known as

A. Fission, where the parent cells divides into roughly two equal halves which mature and become fully functional cells.
B. Budding, where a small part of the cell grows out from the parent cell and eventually breaks free.
C. Spermatogenesis, where the parent cells splits into two daughter cells.
D. Meiosis, where the 2N parent cells splits into four 1N daughter cells.

_____15. The process whereby a portion of the adult living creature can be provided conditions for it to gain structures enabling it to begin living on its own, as found in strawberries, is known as

A. Fission
B. Binary fission
C. Sexual reproduction
D. Vegetative propagation

_____16. Ferns reproduce by producing spores. Spores are considered to be

A. Haploid (2N)
B. Haploid (1N)
C. Diploid (1N)
D. Diploid (2N)

_____17. In ferns, once the spore is released and finds itself in suitable growing conditions it becomes the

A. Gametophyte stage of its life cycle where it remains 1N until finding another 1N individual
B. Gametophyte stage of its life cycle where it remains 2N
C. Sporophyte stage of its life cycle
D. Sporophyte stage of its life cycle where it remains 1N

NAME_____DATE_____

FRIENDLY BIOLOGY

LESSON 13 TEST

Read each question below carefully. Choose the one best answer. Write the letter in the space provided. There may be some responses which refer to more than one response being correct. Read carefully.   An   answer   with   a   misspelled   response   is   a   poor   choice.

_____1. Features or characteristics of an individual are inherited from one's parents. The study of how these features are assorted is known as

A. Taxonomy
B. Genetics
C. Biology
D. Geniology

_____2. Which definition below best fits the term phenotype?

A. Phenotype is what we can observe in an individual based upon its genetic makeup.
B. Phenotype is the record of genes in an individual.
C. Phenotype is how one's genes are arranged on one's chromosomes.
D. Phenotype is the result of years of study on an individual's genotype.

_____3. Variations in the phenotype of a particular gene are known as

A. Genetic alternations
B. Alleles
C. Genetic markers
D. Genotype

_____4. The genetic law of segregation says that

A. During the formation of gametes, the pairs of chromosomes separate with each gamete receiving one or the other member of the pair.
B. All gametes formed during meiosis have the exact same genetic makeup.
C. Only the surviving gametes have the same genetic makeup.
D. Some gametes will be 2N while others will be 1N.

_____5. A second law was discovered by Gregor Mendel which stated that traits are inherited independently from one another. This law is known as the

A. Law of segregation
B. Law of dependent assortment
C. Law in independent assortment
D. Law of implementation

_____6. When various alleles for genes are studied, it is found that some alleles are always expressed regardless if another allele is present from the opposite parent. These types of alleles are said to be

A. Recessive
B. Excessive
C. Dominant
D. Coercive

_____7. When writing letters representing alleles for an individual

A. Dominant alleles are written using upper case letters.
B. Recessive alleles are written using upper case letters.
C. Dominant alleles are written using italics letters.
D. Recessive alleles are written using bold letters.

_____8. Suppose a person had the genotype: HH. This person could be described as being

A. Homozygous dominant
B. Heterozygous dominant
C. Homozygous recessive
D. Heterozygous recessive

_____9. Suppose Tom had the genotype HH. What are the possibilities for the genotypes for his gametes?

A. 50% H 50% h
B. 100% H
C. 100% h
D. 25% H 75% h

_____10. Suppose Tom was married to Susan. If Tom had the genotype HH and Susan had the genotype Hh, would they ever have children with blue eyes? H is the allele for brown eyes while h is the allele for blue eyes.

A. No, they would not.
B. Yes, but only very rarely.
C. Yes, there is an equal chance for Tom and Susan to have brown-eyed children and blue-eyed children.
D. Based on these alleles, they would not be able to have children.

_____11. Suppose Tom was married to Susan. If Tom had the genotype Hh and Susan had the genotype hh, select the statement below which best describes the possible eye colors for their children. Assume H codes brown eyes while h codes for blue eyes.

A. All of their children will have brown eyes.
B. All of their children will have blue eyes.
C. The probability exists that half their children will have blue eyes and half will have brown eyes.
D. The probability exists that 1 out of 4 of their children will have blue eyes.

_____12. Suppose Tom was married to Susan. If Tom had the genotype hh and Susan had the genotype hh, which statement below best describes the possibilities of eye color for their children? Assume H codes for brown eyes while h codes for blue eyes.

A. All of their children will have blue eyes.
B. Most of their children will have blue eyes, however one or two may have brown eyes.
C. All of their children will have brown eyes.
D. Half of their children are likely to have brown eyes and half are likely to have blue eyes.

_____13. Some visible traits have been found to be linked to the sex of the individual as seen in baby chicks. This phenomenon is known as

A. Sex-linkage
B. Sex aligned alleles
C. Gender associated alleles
D. Baby chick gender alleles

_____14. GMO is the abbreviation for

A. Genetically manipulated organism
B. Gene modified organs
C. Genetically modified organism
D. Gene mediated organ

_____15. GMO's may be beneficial for mankind, but may have yet to be discovered disadvantages. Therefore, great care must be taken in development of these organisms.

A. True
B. False

_____16. Men and women each carry a pair of sex chromosomes. Men carry the _____ genotype while women carry the _____ genotype.

A. XX XY
B. XY XX
C. XYZ ZYX
D. AB   CD

_____17. Human sperm can have which possible genotype?

A. All Y
B. Half X and Half Y
C. All X
D. Half A and Half B

_____18. Human ova can have which possible genotype?
A. All Y
B. Half X and Half Y
C. All X
D. Half C and Half D

LESSON 14 TEST

Read each question below carefully. Choose the one best answer. Write the letter in the space provided. There may be some responses which refer to more than one response being correct. Read carefully. An answer with a misspelled response is a poor choice.

_____1. Morphology is the study of

A. The form of living things, mainly how creatures are similar and different from each other.
B. How living things change from one form to another.
C. How living things adapt from one form to another due to seasonal changes.
D. How living things develop and grow over time.

_____2. Taxonomy is the study of

A. The form of living things, mainly how creatures are similar and different from each other.
B. Grouping of living things based upon their morphology.
C. How living things reproduce.
D. How living things have changed.

_____3. The overall main purpose in continual study of taxonomy is

A. Improved sense of how living things reproduce.
B. Better capability to preserve endangered living things.
C. Continued capability to communicate about living things between those studying them.
D. Increased efficiency in public transportation.

_____4. The taxonomist who achieved marked success in classification of many, many living things was

A. George Lucas
B. Carl Linnaeus
C. Gregor Mendel
D. Antonio Avogadro

_____5. A huge database of over one million living creatures exists to allow persons to study taxonomic classification. This database is known as the

A. Catalogue of Life
B. Encyclopedia of Living Things
C. Index of Creatures Known to Mankind
D. Wikipedia of Tomorrow

_____6. There are five main groups that living things are divided into. These five main groups are known as the five

A. Indices
B. Phyla
C. Kingdoms
D. Groups

_____7. Kingdoms are divided into groups known as

A. Classes
B. Phyla
C. Orders
D. Species

_____8. One phylum found in the Animalia kingdom is Chordata. Members of the phylum Chordata all share the common feature of having

A. Feet with two toes
B. Hardened bones
C. A notochord which develops into a spinal column
D. Flippers or fins

_____9. The taxonomic grouping which falls directly under phylum is

A. Class
B. Family
C. Genus
D. Kingdom

_____10. The taxonomic class Mammalia indicates that these animals are alike in that they

A. Are warm-blooded, provide milk for their babies and have fur.
B. Are cold-blooded, lay eggs and have scales.
C. Are found only in the coldest regions of the earth and have fur.
D. Are warm blooded and lay eggs.

_____11. When one writes the scientific name for a living thing, there are conventional rules that are followed. These rules include:

A. The genus name is always capitalized, however the species name is written in lower case letters.
B. The genus name is always underlined, but the species name is not underlined.
C. The species name is always capitalized but the species name is underlined.
D. Both the genus and species names are capitalized and written in italics and also underlined when printed in a textbook.

_____12. Consider this living creature: *Panthera leo (Linnaeus, 1758)*. Which statement below would be <u>incorrect</u> regarding this creature?

A. The genus name for this creature is *Panthera*.
B. The taxonomist who classified this creature was Linnaeus.
C. The year this living creature was classified was 1958.
D. The species name for this living creature is *leo*.

FRIENDLY BIOLOGY

LESSON 15 TEST

Read each question below carefully. Choose the one best answer. Write the letter in the space provided. There may be some responses which refer to more than one response being correct. Read carefully. An answer with a misspelled response is a poor choice.

_____1. Kingdom Animalia includes the

A. Zinnia

B. Snail

C. Bacteria

D. Rabies virus

_____2. Tapeworms are members of the

A. Class Cestoda

B. Class Platyhelminthes

C. Class Annelida

D. United Tapeworm Workers Union

_____ 3. This group of organisms is characterized by having namatocysts

A. Phylum Annelida

B. Phylum Cnidaria

C. Phylum Vertebrata

D. Phylum Chordata

_____4. Flukes are a member of this class.

A. Class Gastropoda

B. Class Trematoda

C. Class Cnidaria

D. Class Mammalia

_____5. Phylum Annelida includes the

A. Segmented worms

B. Flat worms

C. Smelly worms

D. Centipedes

_____ 6. Snails belong to the

A. Class Gastropoda

B. Class Insecta

C. Class Arachnida

D. Class Vertebrata

_____7. Animals with segmented bodies and six legs belong to the

A. Class Insecta

B. Class Arachnida

C. Class Arthopoda

D. Phylum Vertebrata

_____8. A black widow spider would belong to the

A. Class Insecta

B. Class Arachnida.

C. Phylum Arthropoda

D. Class Cnidaria

_____9. Phylum Arthropoda would include the

A. True jellyfish

B. Animals with stinging cells

C. Jointed appendages (legs)

D. Mammals

Phylum Chordata includes

A. Microorganisms

B. Animals with backbones

C. Animals without backbones

D. All animals

NAME_____DATE_____
FRIENDLY BIOLOGY
LESSON 16 TEST

Read each question below carefully. Choose the one best answer. Write the letter in the space provided. There may be some responses which refer to more than one response being correct. Read carefully. An answer with a misspelled response is a poor choice.

_____1. Class Osteichthyes includes the
A. Bony fishes
B. Cartilagenous fishes
C. Snakes
D. Reptiles

_____2. Order Anura includes the
A. Frogs
B. Salamanders
C. Bony fishes
D. Monkeys and lemurs

_____3. Animals which spend part of the life cycle in water and then part on land are found in the
A. Class Anura
B. Class Amphibia
C. Class Osteichtheses
D. Class Mammalia

_____4. Pigeons, blue jays and sparrows would all be found in
A. Class Aves
B. Class Apoda
C. Class Mammalia
D. Class Amphibia

_____5. Class Reptilia includes
A. A. Tapeworms
B. Lizards
C. Mongooses
D. Sea Cows

_____6. In the Order Rhynchocephalia we would likely find
A. Tuatara
B. Seals
C. Porpoises
D. The family living next door

_____7. Anacondas, rattlesnakes and garter snakes belong to the
A. Order Squamata
B. Order Chiroptera
C. Order Rhynchocephalia
D. Class Mammalia

_____8. Order Chiroptera includes animals that
A. Have scales
B. Have feathers and wings
C. Eat only ocean creatures
D. Utilize echolocation

_____9. Order Crocodilia includes animals which are
A. Carnivores
B. Herbivores
C. Omnivores
D. Carnavores

_____10. Class Mammalia includes animals which
A. Provide milk to their offspring
B. Have jointed legs and antennae
C. Are only found in marine environments
D. Lay eggs

_____21. Mice, rats and squirrels belong to the
A. Order Mammalia
B. Order Rodentia
C. Class Anura
D. Class Apoda

_____23. Order Marsupialis are those animals which
A. Give birth and then quickly abandon all but one of their offspring.
B. Give birth and carry their offspring in a pouch.
C. Have flattened scales and lay eggs.
D. Are active only at night and utilize echolocation.

NAME_____DATE_____

FRIENDLY BIOLOGY

LESSON 17 TEST

Read each question below carefully. Choose the one best answer. Write the letter in the space provided. There may be some responses which refer to more than one response being correct. Read carefully. An answer with a misspelled response is a poor choice.

_____1. Plants, through the process of _____ are able to take carbon dioxide, water and sunlight to produce _____ which is fuel for the plant.
A. Photosynthesis, a lipid
B. Photosynthesis, glucose
C. Respiration, glucose
D. Hydrolysis, a protein

_____2. Plants are considered to be _____ in that they are able to produce their own food supply.
A. Omnivores
B. Herbivores
C. Autotrophs
D. Carnitrophs

_____3. Photosynthesis in plants takes place in the _____ of the cells of plants.
A. Cell walls
B. Nucleus
C. Ribosomes
D. Chloroplasts

_____4. Plants have cell walls which are made up primarily of _____ which are long chains of glucose molecules.
A. Gasoline
B. Sucrose
C. Cellulose
D. Cellulase

_____5. Taxonomists divide plants into two large groups based upon the presence of
A. Roots and stems
B. Vascular tissue
C. Type of chlorophyll present
D. Shape of leaves

6. Phloem is the vascular tissue of plants which carries
A. Water
B. Food like glucose

C. Insects

D. Decaying vegetable matter

7. Trees such as firs, spruces and pines have seeds that form in cones and belong to the Division

A. Gingkophyta

B. Anthophyta

C. Monocotophyta

D. Coniferophyta

8. Plants which have flowers and seeds enclosed in a fruit belong to the Division

A. Gingkophyta

B. Anthophyta

C. Monocotophyta

D. Coniferophyta

9. There are two classes found within the Division Anthophyta. The first includes plants with one cotyledon and are commonly referred to as

A. Monocots

B. Bicots

C. Dicots

D. Camping cots

10. Beans and other legumes are plants that have two cotyledons. They belong to the class

A. Monocotyledonae

B. Dicotyledonae

C. Bicotyledonae

D. Tricotyleconae

11. The male parts of a flower are the stem-like _____ with the pollen-producing part known as the _____.

A. Anthers, stigmas

B. Pistils, anthers

C. Anthers, gametophytes

D. Filaments, anthers

12. Within the ovary of a flower, one finds the

A. Pollen granules

B. Developing sperm cells

C. Ovules

D. New flower buds

LESSON 18 TEST

Read each question below carefully. Choose the one best answer. Write the letter in the space provided. There may be some responses which refer to more than one response being correct. Read carefully. An answer with a misspelled response is a poor choice.

_____1. Members of the Kingdom Monera are considered to be prokaryotic meaning they
A. Have no cell membranes present
B. Have cells walls with cellulose like plants
C. Have no nuclear membranes nor organized organelles
D. Are totally autotrophic

_____2. Members of the Kingdom Monera also are _____ meaning they solely consist of being one cell.
A. Heterophilic
B. Unicellular
C. Unicycular
D. Eukaryotic

_____3. The primary forms of reproduction in the Kingdom Monera is
A. Sexual and asexual reproduction.
B. Fission and budding
C. Vegetative propagation and fragmentation
D. Sporophyte generation

4. The phylum of Monera which includes the bacteria which causes disease is the Phylum
A. Schizophyta
B. Mammalia
C. Archaebacteria
D. Holophilia

5. Viruses consist of a strand of _____ or _____ enclosed by a protein coat known as the capsid
A. RNA or tRNA
B. DNA or RNA
C. ABC or DEF
D. mRNA or mDNA

6. Viruses are considered to be obligate intracellular agents as they must locate within the cells of a host to survive.
A. True
B. False

7. Control of viruses is mainly through the use of _____ which stimulate the body's immune system to produce _____ to fight off the disease when encountered at a later time.

A. Large clubs, antibiotics
B. Antibiotics, antibodies
C. Vaccines, antibiotics
D. Vaccines, antibodies

8. Viruses work by entering a host's cells and taking over control of the cell's _____ in an effort to get the cell to make more viruses. The host cell then dies.

A. Rough ER
B. Lysosomes
C. DNA
D. Microtubules

9. Viruses are classified as either being _____ or _____ viruses.

A. Good, bad
B. DNA or TNA
C. RNA or DNA
D. Small or smaller

10. Control of viruses through the use of antibiotics is usually successful.

A. True
B. False

FRIENDLY BIOLOGY

LESSON 19 TEST

Read each question below carefully. Choose the one best answer. Write the letter in the space provided. There may be some responses which refer to more than one response being correct. Read
carefully. An answer with a misspelled response is a poor choice.

1. Members of the Kingdom Protista have membrane-bound organelles and are therefore said to be
A. Prokaryotic
B. Eukaryotic
C. Abiotic
D. Anucleated

2. Members of the animal-like protistas are grouped based upon _____.
A. Means of locomotion
B. Color
C. Size
D. Flavor

3. _____ are the tiny hair-like projections found in the Phylum Cilophora.
A. Cilia
B. Silia
C. Flagella
D. Pseudopoida

4. Members of the Phylum Zoomastigina move about through the use of a _____ which is a tail
-like structure capable of propelling the organism through the water.
A. Cilium
B. Flagellum
C. Hair
D. Pseudopodia

5. The fungal-like protistas include slime molds which move about on the _____
eating up dead leaves.
A. Ocean floor
B. Forest floor
C. Kitchen floor
D. Desert floor

6. The devastating _____ famine of the 1800's was caused by a member of
the Oomycota phylum.
A. Tomato

B. Potato

C. Fish

D. Barley

7. The plant-like protistas include various phyla of _____.

A. Algae

B. Monocots

C. Cotyledons

D. Ciliates

8. Algae are categorized based upon the presence of various colors of
_____ as well as how they store _____.

A. Pigments, lipids

B. Pigments, glucose

C. Mitochondria, glucose

D. Cellulose, proteins

9. Chlorophyta are the _____ algae.

A. Green

B. Red

C. Blue

D. Brown

10. Rhodophyta are the _____ algae.

A. Green

B. Red

C. Blue

D. Brown

FRIENDLY BIOLOGY
LESSON 20 TEST

Read each question below carefully. Choose the one best answer. Write the letter in the space provided. There may be some responses which refer to more than one response being correct. Read carefully. An answer with a misspelled response is a poor choice.

1. Fungi are _____meaning they live off of dead organisms.
A. Carnivores
B. Saprophytes
C. Herbivores
D. Parasites

2. Fungi are categorized according to the presence or lack of _____ in their hyphae.
A. Crosswalls

B. Holes

C. Branches

D. Nuclei

3. Fungi can reproduce sexually and asexually.
A. True

B. False

4. Fungi which belong to the Division _____ have no crosswalls in their hyphae and the common black bread mold is a member.
A. Zygomycota

B. Ascomycota

C. Basidiomycota

D. Ilikeyournewcota

5. Fungi which belong to the Division _____ have crosswalls in their hyphae. Familiar members include mushrooms you might serve on a pizza.
A. Zygomycota

B. Basidiomycota

C. Chlorphycota

D. Achlorphycota

6. Morels are members of the fungal division

A. Deuteromycota

B. Basidiomycota

C. Ascomycota

D. Zygomycota

7. The fungal organism which causes athlete's foot belongs to the division

A. Deuteromycota

B. Basidiomycota

C. Ascomycota

D. Zygomycota

FRIENDLY BIOLOGY

LESSON 21 TEST

Read each question below carefully. Choose the one best answer. Write the letter in the space provided. There may be some responses which refer to more than one response being correct. Read carefully. An answer with a misspelled response is a poor choice.

1. The study of how living things are made is known as
A. Biology
B. Histology
C. Anatomy
D. Physiology

2. The study of how living things function is known as
A. Cytology
B. Histology
C. Anatomy
D. Physiology

3. The hardening process of the bones is called
A. Osteology
B. Ossification
C. Osteocytes
D. Osteoblasm

4. The boundary where growth of the bone occurs is referred to as the
A. Growth plate
B. Growth chasm
C. Synovial plate
D. Diaphysis

5. The joints between of the long bones of the arms and legs are known as
A. Sutures
B. Ligaments
C. Tendons
D. Synovial joints

6. If you look at the ends of long bones, you'll find a layer of _____cartilage.
A. Hyaline
B. Epiphyseal
C. Black
D. Rough

7. The fluid to lubricate synovial joints is known as
A. Joint elixir
B. Articular lubricant
C. Synovial fluid
D. Fluid de la articula

8. There are two kinds of osteocytes. The _____ are bone cells which have the responsibility of building new bone, like along the growth plate, while the _____ have the job of taking apart or tearing-down bone.
A. Osteoblasts, osteoclasts
B. Osteoclasts, osteoblasts

9. Besides the function of movement, bones also function to
A. Protect internal organs and store bone marrow and fatty tissue
B. Maintain levels of water in the body
C. Provide a source of bile salts necessary for digestion to take place
D. Produce white blood cells as needed

10. Skeletal muscle is also referred to as _____ muscle.
A. Smooth
B. Cardiac
C. Striated
D. Non-smooth

11. The two types of contractile proteins are:
A. A and B
B. DNA and RNA
C. Actin and myosin
D. Alpha and omega

12. The attachment of muscles to bones is made through tough connective tissue material known as
A. Tendons
B. Ligaments
C. Joint capsules
D. Cartilage

13. Connective tissue structures which connect bones to bones are known as

A. Tendons

B. Ligaments

C. Marrow

14. Bending of a joint is referred to as flexion while straightening a joint is known as extension.

A. True

B. False

15. Cells of the nervous systems are called

A. Erythrocytes

B. Neurons

C. Osteocytes

D. Osteoblasts

16. A single dendrite leaves a nerve cell while there are multiple axons.

A. True

B. False

17. At the terminal end of a neuron there are vesicles containing _____ which when released trigger a muscle to contract.

A. Nerve secretions

B. Axonal excretions

C. Neurotransmitters

D. Elevated ions

18. Neurons which carry messages out to muscles are known as _____ or _____neurons.

A. Afferent, sensory

B. Efferent, motor

19. Neurons which carry messages inward from sensory organs such as the skin or eyes are called _____ or _____ neurons.

A. Afferent, sensory

B. Efferent, motor

LESSON 22 TEST

Read each question below carefully. Choose the one best answer. Write the letter in the space provided. There may be some responses which refer to more than one response being correct. Read carefully. An answer with a misspelled response is a poor choice.

1. The primary function of the _____ system is to deliver oxygen and glucose to all cells of the body.
A. Respiratory
B. Circulatory
C. Digestive
D. Renal

2. The smallest blood vessels are known as the
A. Capillaries
B. Arterioles
C. Venules
D. Arteries

3. Openings within the walls of capillaries are known as
A. Arterations
B. Fenestrations
C. Hyphae
D. Septae

4. In the human heart there are four _____ .
A. Chambers
B. Arteries
C. Ventricles
D. Atria

5. The heart chamber which receives blood low in oxygen from the body is the
A. Left ventricle
B. Left atrium
C. Right ventricle
D. Right atrium

6. Blood passes through a valve as it moves from the right atrium to the right ventricle. This valve is the:
A. Right AV valve
B. Left AV valve

C. Left semilunar valve

D. Pyloric valve

7. Blood entering the left atrium is high in oxygen and low in carbon dioxide.

A. True

B. False

8. The right and left atrioventricular valves consist of flap-like structures known as the _____ of the valves.

A. Gates

B. Buttons

C. Cusps

D. Cuffs

9. Chordae tendonae prevent a valve from

A. Enlarging

B. Prolapsing

C. Shrinking

D. Flopping

10. Valves that do not close correctly or those that don't open wide enough may produce abnormal heart sounds. These sounds are known as

A. Heart songs

B. Heart mimics

C. Heart acoustics

D. Heart murmurs

11. The trachea branches into bronchi which branch into bronchioles with end in alveoli where gases are exchange between the capillaries and the air space.

A. True

B. False

12. The watery portion of blood is known as

A. Plasma

B. Hydrolysis

C. Leukocytes

D. Neutrophils

13. Erythrocytes are produced in the bone marrow and have the primary job of

A. Fighting disease

B. Carrying oxygen

C. Defending invaders

D. Allowing growth to occur

14. Hemoglobin is
A. A protein used to fight disease
B. A lipid used to carry oxygen
C. A protein used to carry oxygen
D. What gives the blood its dark blue color

15. The decrease in the functioning number of red blood cells is known as
A. Hypercythemia
B. Anoxia
C. Anemia
D. Bilirubinemia

16. The most abundant white blood cell which has a polymorphic nucleus is the
A. Neutrophil
B. Basophil
C. Lymphocyte
D. Platelet

17. Lymphocytes create
A. Antibiotics to fight disease
B. Antibodies to attack invaders
C. Inflammation due to the attack of an invader
D. Beautiful works of art

LESSON 23 TEST

Read each question below carefully. Choose the one best answer. Write the letter in the space provided. There may be some responses which refer to more than one response being correct. Read carefully. An answer with a misspelled response is a poor choice.

1. The main function of the digestive system is to
A. Take food and convert it into glucose to be delivered to the body.
B. Remove toxins from the food we eat in order for us to live healthier lives.
C. Pump blood throughout the body.
D. Open the doors of the cells in order for glucose to enter.

2. A bite of food is known as a
A. Codling
B. Bolus
C. Nolus
D. Gulp

3. Secretions from salivary gland contain the enzyme _____ to being the digestion process.
A. Cellulose
B. Cellulase
C. Amylose
D. Amylase

4. The tube which transports food from the mouth to the stomach is the _____ and consists of _____ muscles.
A. Esophagus, striated
B. Esophagus, smooth
C. Duodenum, smooth
D. Jejunum, rough

5. Two types of cells are found in the lining of the stomach. The parietal cells produce
A. Perchloric acid
B. Arachidonic acid
C. Sulfuric acid
D. Hydrochloric acid

6. The stomach has two "doors." The front door is the _____ sphincter while the back door is the _____ sphincter.
A. Cardiac, pyloric
B. Pyloric, anal
C. Anal, cardiac
D. Cardiac, anal

7. The longest section of small intestine whose main job is absorption of nutrients is the
A. Duodenum
B. Jejunum
C. Ileum
D. Colon

8. The pancreas secretes the hormone _____ which regulates the ability of _____ to enter cells of the body.
A. Glucagon, glucose
B. Insulin, heptose
C. Epinephrine, lactose
D. Insulin, glucose

9. The liver has an incredible number of jobs it does for the body. One job is the creation of glucagon from glucose. Glucagon is
A. An enzyme which readily lyses protein.
B. The storage form of glucose.
C. An intermediate metabolite in the respiratory process.
D. The glue used to cement cells together.

10. The two main roles of the large intestine is
A. Water production and Vitamin B assimilation
B. Water conservation and Vitamin K production
C. Nutrient absorption and Vitamin A production
D. Nutrient absorption and Vitamin K production.

FRIENDLY BIOLOGY

LESSON 24 TEST

Read each question below carefully. Choose the one best answer. Write the letter in the space provided. There may be some responses which refer to more than one response being correct. Read carefully. An answer with a misspelled response is a poor choice.

1. The systems of our body work to remove wastes created by cells as well as cell remnants from the body are the
A. Digestive and renal systems
B. Renal and urinary systems
C. Reproductive and urinary systems
D. Circulatory and nervous systems

2. The "normal" human has _____ kidney(s.)
A. One
B. Two
C. Three
D. Four

3. The filtration unit of the kidney is
A. Glomerulus
B. Renal pelvis
C. Ureter
D. Urethra

4. In addition to blood filtration, the kidneys also maintain _____ of the blood.
A. Color
B. pH
C. Temperature
D. Mass

5. The kidneys also produce hormones which control blood pressure and the creation of
A. White blood cells
B. Red blood cells
C. Gametes
D. Neurons

6. The presence of protein in the urine indicates possible kidney damage or damage to other parts of the urinary system.
A. True
B. False

7. Glucose in the urine is a normal finding and should not be of concern.
A. True
B. False

8. Lack of insulin can result in _____ levels of blood glucose, presence of glucose in the _____ and "starving" cells all over the body.
A. High, urine
B. Low, urine
C. Normal, urine
D. High, feces

9. Urine leaves the kidneys through the _____ which drain into the _____.
A. Bladder, kidneys
B. Urethra, bladder
C. Ureters, urethra
D. Ureters, bladder

10. When full, the bladder is emptied to the exterior of the body through the
A. Ureter
B. Afferent arteriole
C. Vas deferens
D. Urethra

NAME_____DATE_____
FRIENDLY BIOLOGY

LESSON 25 TEST

Read each question below carefully. Choose the one best answer. Write the letter in the space provided. There may be some responses which refer to more than one response being correct. Read carefully. An answer with a misspelled response is a poor choice.

1. Glands which produce substances which are released directly into the circulatory system and not through a tube or duct are known as
A. Endocrine glands
B. Exocrine glands
C. Salivary glands
D. Prostate glands

2. Endocrine glands in the body regulate levels of body hormones by being stimulated when hormone levels fall to certain levels. This system of control is known as a
A. Positive reflex system
B. Negative uptake system
C. Negative feedback mechanism
D. Neutral inhibition system

3. The _____ is known as the "master" gland.
A. Adrenal
B. Pituitary
C. Cowper's
D. Kidney

4. TSH affects rate of _____ which is how fast or slow cells utilize _____.
A. Digestion, ribose
B. Urination, glucose
C. Metabolism, glucose
D. Respiration, sucrose

5. ACTH signals the _____to produce anti-inflammatory substances and water conservation hormones.
A. Adrenal glands
B. Kidneys
C. Bladder
D. Testis

6. HGH regulates _____ in the body.
A. Sexual activity
B. Metabolism
C. Growth
D. Inflammation

7. Prolactin has its effects upon
A. Milk producing tissues
B. Bone cells

C. Reproductive organs
D. Neurons

8. The gonadotropic hormones have their effects upon the
A. Ovaries in women and testis in men.
B. Kidneys
C. Adrenal glands in women and ovaries in men.
D. Thyroid glands

9. FSH stimulates the ovary in women to produce
A. Follicles with an ovum inside
B. Higher levels of calcium in the body
C. Male gametes
D. Spores for 1N meiosis

10. Oxytocin has its effects at two different locations in women. Those locations include the
A. Uterus and milk producing glands.
B. Ovaries and thyroid glands
C. Breast tissue and growth area of bones
D. Adrenal glands and parathyroid glands

11. The parathyroid glands control _____ levels in the body.
A. Potassium
B. Calcium
C. Carbon
D. Sodium

12. The _____ gland regulates daily sleep and wakefulness patterns.
A. Melatonin
B. Pineal
C. Salivary
D. Pituitary

NAME_____DATE_____
FRIENDLY BIOLOGY

Read each question below carefully. Choose the one best answer. Write the letter in the space provided. There may be some responses which refer to more than one response being correct. Read carefully. An answer with a misspelled response is a poor choice.

1. Which statement below best describes ova?
A. Ova are 2N cells produced by the ovaries.
B. Ova are 1N cells produced by the testis.
C. Ova are haploid cells produced by the ovaries.
D. Ova are diploid cells produced by the testis.

2. The testis are located outside the body in the
A. Placenta
B. Uterus
C. Scrotum
D. Abdomen

3. The ovaries are located in the
A. Scrotum
B. Abdomen
C. Chest
D. Neck

4. Choose the best description for the pathway sperm cells follow to exit the body.
A. Testis, epididymis, vas deferens, urethra
B. Testis, vas deferens epididymis, urethra
C. Testis, urethra, epididymis, vas deferens
D. Testis, vas deferens, epididymis, urethra

5. The accessory sex glands in the male provide nutritional support and pH adjusting substances to enable the sperm to survive.
A. True
B. False

6. The cervix functions like a door to open at specific times in response to hormones to allow sperm to enter or be closed during pregnancy.
A. True
B. False

7. The release of the ovum is known as
A. Ovulation
B. Fertilization
C. Follicle stimulation
D. Cultivation

8. Fertilization usually takes place in the
A. Uterus

B. Fallopian tubes
C. Cervix
D. Testis

9. After fertilization, the embryo moves down to the _____ to implant and continue development.
A. Urethra
B. Ureter
C. Uterus
D. Prostate gland

10. Blood never mixes directly between the mother and developing baby.
A. True
B. False

11. _____ from the posterior pituitary gland stimulates the uterus to begin contractions while decreasing _____ levels allows the cervix to dilate.
A. FSH, estrogen
B. Oxytocin, progesterone
C. Estrogen, GNRH
D. Testosterone, TSH

12. Should the baby be too large to deliver, a _____ may be performed to surgically remove the baby from the uterus.
A. Cesarean section
B. Hysterectomy
C. Global laparotomy
D. Abdominal suture

NAME_____DATE_____
FRIENDLY BIOLOGY
LESSON 27 TEST

Read each question below carefully. Choose the one best answer. Write the letter in the space provided. There may be some responses which refer to more than one response being correct. Read carefully. An answer with a misspelled response is a poor choice.

1. The clear outer covering of our eyes is called
A. Lens
B. Sclera
C. Cornea
D. Iris

2. This gland produces tears to maintain adequate moisture to the surface of the eye.
A. Adrenal gland
B. Lacrimal or tear gland
C. Salivary gland
D. Endocrine gland

3. The colored portion within the eye is the _____ and creates the circular _____ which adjusts the amount of light which can enter the eye.
A. Pupil, lens
B. Iris, retina
C. Iris, pupil
D. Cornea, pupil

4. The fluid in front of the lens of the eye is the aqueous humor while the fluid behind the lens, which is much thicker, is the vitreous humor.
A. True
B. False

5. Excess fluid within the eye is
A. Diabetes
B. Glaucoma
C. Retinitis
D. Conjunctivitis

6. Tiny muscles encircling the lens allow it to change shape which in turn allows images to be focused on the
A. Iris
B. Retina
C. Auditory canal
D. Sclera

7. The cranial nerve which takes nerve stimuli from the retina to the brain is the
A. Optic nerve
B. Auditory nerve
C. Accessory nerve
D. Abducence nerve

8. Sound moves into the auditory canal and strikes the _____ membrane which causes the three tiny bones of the middle ear to vibrate.
A. Cochlea
B. Auricular
C. Tympanic
D. Auditory

9. Vibrations from the malleus, incus and stapes stimulate nerve endings within the fluid-filled cochlea which then send signals by the optic nerve to the brain.
A. True
B. False

10. The semicircular canals within the ear work to maintain
A. Happiness
B. Balance
C. Good relationships
D. Financial security

11. Sensations of smell travel from nerve endings in the nose to the brain along the
A. Auditory nerve
B. Optic nerve
C. Eustacian nerve
D. Olfactory nerve

12. On the surface of the tongue one finds _____ which house taste buds which consist of taste receptor cells.
A. Papillae
B. Olfactory receptors
C. Lingual fissures
D. Tympanic membranes

13. The skin consists of two layers: the dermis which creates squamous cells that form the surface to the skin and the deeper epidermis which houses blood vessels, nerves, glands and fatty tissue.
A. True
B. False

14. Touch sensations travel by _____ neurons to the spinal cord and brain.
A. Afferent or sensory
B. Efferent or motor

15. The skin can be divided into regions of sensation. These regions are known as
A. Skin tags
B. Dermatomes
C. Sweat glands
D. Dermal zones

LESSON 28 TEST

Read each question below carefully. Choose the one best answer. Write the letter in the space provided. There may be some responses which refer to more than one response being correct. Read carefully. An answer with a misspelled response is a poor choice.

1. Ecology is the study of the relationships between living things and
A. Other living things.
B. Their environment.
C. Humans
D. Their food supplies

2. Which statement below best describes a biome?
A. An area or region of the earth where similar organisms thrive and others do not.
B. A small area with a diverse number of living organisms present.
C. A portion of the earth where no living things are present.
D. A portion of the earth where members of all regions of the earth are naturally present.

3. This biome has the coldest average temperatures with very little plant or animal life present.
A. Polar biome
B. Grassland biome
C. Tundra
D. Desert

4. Very short growing plants and a variety of animals can be found in this biome. It is frequently observed on mountain peaks and in regions of the earth bordering the polar biomes.
A. Tundra
B. Desert
C. Temperate rainforest
D. Grassland savannah

5. Spruces, firs and pines are often found in this biome.
A. Coniferous biome
B. Grassland biome
C. Tropical rainforest
D. Deciduous forest

6. Trees such as maples, oaks and hickories are often found in this biome.
A. Deciduous forest
B. Coniferous forest
C. Grassland
D. Tropical rainforest

7. Minimal precipitation and warm temperatures are indicative of the _____ biome.
A. Deciduous grassland
B. Desert
C. Temperate rainforest
D. Tundra

8. The temperate rainforest biome receives great amounts of rainfall annually but does not have as warm temperatures as the tropical rainforest.
A. True
B. False

9. The _____ biome consists of the earth's oceans and seas.
A. Polar
B. Marine
C. Aquatic
D. Tropical

10. The freshwater biome consists of lakes, rivers and streams and the location where the marine and freshwater biomes meet is known as an estuary.
A. True
B. False

11. The organisms which are capable of capturing energy from the sun and converting it into useable food for itself (and later, others) are known as
A. Heterotrophs or consumers
B. Autographs or producers
C. Autotrophs or producers
D. Herbivores or consumers

12. Heterotrophs "harvest" energy captured from the sun through the functions of
A. Plants (producers.)
B. Carnivores (consumers.)
C. Consumers
D. Indicative members

13. The consumer which first eats a producer (plant) is known as a _____ consumer.
A. Primary
B. Elementary
C. Basal
D. Heterotrophic

14. Organisms which eat a primary consumer are called _____ consumers.
A. Elementary
B. High School
C. Secondary
D. College

15. _____ are those organisms which consume dead organisms.
A. Predators
B. Herbivores
C. Scavengers
D. Carnivores

16. With each step through an ecosystem away from the producer level, the amount of available energy is decreased.
A. True
B. False

TESTS 1-14 ANSWER KEY

QUESTION #	1	2	3	4	5	6	7	8	9	10	11	12	13	14
1	B	A	B	B	B	B	C	C	D	B	B	B	B	A
2	A	D	C	C	B	B	C	D	B	C	C	B	A	B
3	B	B	A	C	A	C	B	A	D	B	C	B	B	C
4	C	C	B	A	C	D	C	A	A	A	D	C	A	B
5	A	D	D	B	B	B	D	B	A	A	D	A	C	A
6	B	C	A	A	A	B	A	D	C	B	A	B	C	C
7	E	A	C	A	C	D	B	A	C	C	C	A	A	B
8	C	B	B	B	D	B	B	D	A	B	A	C	A	C
9	E	D	B	B	A	A	A	D	B	C	A	C	B	A
10	B	C	D	A	A	B	B	C	C	C	B	B	A	A
11		D	C	D	B	A	A	A		B	A	A	C	A
12		D	B	B	D		C	A			A	D	A	C
13			C	D	D		B					B	A	
14			B				C					A	C	
15			A									D	A	
16			C									B	B	
17			A									A	B	
18			D										C	
19			C											
20			A											
21			B											
22			C											

TESTS 15-28 ANSWER KEY

QUESTION #	15	16	17	18	19	20	21	22	23	24	25	26	27	28
1	B	A	B	C	B	B	C	B	A	B	A	C	C	B
2	A	A	C	B	A	A	D	A	B	B	C	C	B	A
3	B	B	D	B	A	A	B	B	D	A	B	B	C	A
4	B	A	C	A	B	A	A	A	B	B	C	A	A	A
5	A	B	B	B	B	B	D	D	D	B	A	A	B	A
6	A	A	B	A	B	C	A	A	A	A	C	A	B	A
7	A	A	D	D	A	A	C	A	B	B	A	A	A	B
8	B	D	B	C	B		A	C	D	A	A	B	C	A
9	C	A	A	C	A		A	B	B	D	A	C	B	B
10	B	A	B	B	B		C	D	B	D	A	A	B	A
11		B	D				C	A			B	B	D	C
12		B	C				A	A			B	A	A	A
13							B	B					B	A
14							A	C					A	C
15							B	C					B	C
16							B	A						A
17							C	B						
18							B							
19							A							
20														
21														
22														

ANSWER KEYS FOR LESSON PRACTICE PAGES

Name _____Date_____

Lesson 1   Practice Page 1

Instructions: Fill-in the blank with the appropriate word. Refer back to the text portion of the lesson for help.

1. The term biology is derived from two words: ___bios_____ which means life and ___ology_____ which means the study of.

2. There are other words like biology. The study of the earth is _____geology_____ while the study of ancient artifacts is _____archeology_____.

3. Hematology is the study of _____blood_____ and the study of tumors or cancers is known as _____oncology_____.

4. Being able to __move____ or change location or position is evidence of life in a living thing.

5. A feature common to all living things is that they can make new living things. This process is known as ___reproduction_____.

6. Living things all require a source of __energy__ in order to carry out life processes. For humans this source is the ___food___ we eat, however for most plants, it is the __sun__ which provides this source.

7. Small living things increase in size and complexity. Living things ___grow____ and develop.

8. Living things react or ____respond___ to things around them which is also known as their environment.

9. __Death___ is the condition where evidence of life no longer exists.

10. The name of the primary author of this textbook is _____Dr. Joey Hajda___.

Read each example below and tell whether which form of evidence of life is being demonstrated. Some examples may have more than one evidence of life present.

Example: A puppy chases a ball. <u>Movement</u>

Tommy is hungry and so he eats three hot dogs for lunch. ____Need for energy source__

11. A tree sparrow flees from an approaching cat. __Movement or responds to environment or need for energy source_____

12. The goose in incubating ten eggs. ____Reproduction____

13. The earthworm wriggles down deeper into the bedding in the bait cup. Responds to environ-

ment_____

15. The sunflower tilts its flower towards the east in the morning hours. _____Movement or responds to environment_____

16. Mushrooms tend to be found most frequently near dead or dying trees. _Need for energy source_____

17. The kitten learns to catch mice by imitating its mother. ___Grow and develop_____

18. As the summer month go by, the ears of corn on the corn plants appear to be getting larger in diameter. ____Grow and develop or reproduction___

19. Allowing food to sit out unrefrigerated and then eating it can result in food poisoning. _____Respond to environment or reproduce or grow and develop_____

20. The peach tree developed pink blossoms as the temperatures increased. ___Respond to environment or reproduce_____

Choose an animal in your home or one you are familiar with. Tell what it is and then give four pieces of evidence that tells you it is a living thing. For example, if you choose your dog, a piece of acceptable evidence is that it needs to be fed everyday.

21._____Answers vary_____

    1.

    2.

    3.

    4.

Choose a plant in your yard or one you are familiar with. Tell what it is and then give four pieces of evidence that tells you it is a living thing.

22._____Answers vary_____

    1.

    2.

    3.

    4.

Name _____Date_____

Lesson 1   Practice Page 2

Instructions: Below you will find clues to solve this crossword puzzle. Refer back to the text portion of the lesson for help.

Across

2. living things __GROW___ and develop

3. process of a living thing making another living thing   REPRODUCE

4. requirement of living things in order to carry out life processes   ENERGY SOURCE

6. when evidence of life no longer is present in an organism DEATH

8. living things _____to their environment RESPOND

10. author of this textbook JOEY HAJDA

11. tumors or cancer ONCO

12. life BIOS

Down

1. changing position or location as indicator of being alive MOVEMENT

5. earth GEOS

7. one complete living thing ORGANISM

9. the study of OLOGY

Name _____Date_____

Lesson 2   Practice Page 1

Instructions: Fill-in the blank with the appropriate word. Refer back to the text portion of the lesson for help.

1. All things, whether living or non-living consist of tiny bits of matter known as _atoms_.

2. The central portion of an atom is known as the ___nucleus__ and it contains subatomic particles known as the ___neutrons_____ and the ___protons_____.

3. Circling around the nucleus of the atom are a third type of subatomic particle which are the ____electrons_____.

4. Theories say it is the _____arrangement__ of the electrons which determines the behavior of various elements on the periodic table.

5. To determine the number of protons or electrons an atom of a particular atom has, one looks for the ____atomic number___ of that element on a periodic table.

6. The electrons are thought to exist in ___orbitals or shells___ around the nucleus of an atom and that there can be no more than ____eight electrons___ on one of these layers.

7. Elements which have their outermost layer of electrons filled are the elements which are very _____stable____ in their behavior.

8. Elements which have their outermost layers incompletely filled are elements which are very _____reactive or unstable_____ in their behavior.

9. Elements which are very reactive seek to gain stability by moving or sharing ____electrons_____ with neighboring atoms of elements.

10. The family of elements whose atoms have their outer layers of electrons completely filled making their very, very stable is the ____noble gas or inert gas____ family.

11. Atomic bonds which form between atoms who have transferred electrons from one to another are known as ___ionic__ bonds.

12. Atomic bonds which form between atoms who are sharing electrons between themselves are known as ____covalent___ bonds.

13. Of the many elements known to man, there are four that are common to all living things and deserve our special attention. Those four elements are: __carbon C ___, ____hydrogen H___, ____oxygen O____ and ____nitrogen N____. Write their element symbols next to their names, too!

14. Of the elements listed below, choose the one that would most likely be the least reactive.

    A. Hydrogen

    B. Carbon

    C. Sodium

    <u>D. Neon</u>

15. Of the elements listed below, choose the one that would most likely be the most reactive.

    A. Neon

    <u>B. Sodium</u>

    C. Argon

    D. Helium

16. Which subatomic particle is thought to be responsible for an atom's behavior?

    A. Proton

    B. Neutron

    <u>C. Electron</u>

    D. Crouton

17. Suppose Atom A desires to get rid of one electron and Atom B is willing to accept that one electron. Together, by moving this electron, they can become a compound which is stable. This type of bond formation where electrons are moved is called a(n)

    A. Proton bond

    B. Single covalent bond

    C. Double covalent bond

    <u>D. Ionic bond</u>

    E. James bond (LOL)

18. There are many elements required by living things in order to maintain life. There are four that we will study next. Circle these four important elements found in this list of element symbols:

<u>H</u>   He   <u>C</u>   Ca   Li   Be   Ar   Ox   <u>O</u>   Br   Ni   <u>N</u>   Zn   As

Pb   Cu

Name _____ Date_____

Lesson 2   Practice Page 2

Instructions: Below you will find clues to solve this crossword puzzle. Refer back to the text portion of the lesson for help.

Across

3. subatomic particle analogous to planets in our solar system   ELECTRONS

6. defined as a change in position or location, a feature of living creatures MOVEMENT

10. It is the _ARRANGEMENT_____ of electrons which determines reactivity of an element.

11. a feature of living things where more living things just like the parents are created REPRODUCE

13. an important element found in living organisms, a component of our air OXYGEN

14. the central part of an atom NUCLEUS

Down

1. the type of bond formed when electrons get transferred from one atom to another IONIC BOND

2. an important element of living organisms which has the atomic number six CARBON

4. tiny bits of matter from which all things are thought to be made ATOMS

5. the type of bond formed when atoms share electrons between themselves to gain stability COVALENT BOND

7. an important element of living organisms which has the symbol N NITROGEN

8. an important element in living organisms, also found in water HYDROGEN

9. electrons are arranged in _____ around the nucleus of an atom LAYERS

12. subatomic particles found in the nucleus of atoms PROTONS AND NEUTRONS

Name _____Date_____

Lesson 3   Practice Page 1

Instructions: Fill in the blank with the appropriate word. Refer back to the text portion of the lesson for help.

1. The common name for carbohydrates is _____sugars_____.

2. The word carbohydrate is derived from two words carbo– which indicates the presence of _____carbon_____ and -hydrate which means _____water_____ is also present.

3. The generic formula for a carbohydrate is _____$C_n(H_2O)_n$_____ where the little letter "n" tells _____the number of water molecules present_____.

4. The carbohydrate in which n = 6 has the formula _____$C_6H_{12}O_6$_____ and its name is _____glucose_____. Its common name is _____blood sugar_____,

5. Another carbohydrate in which n = 6 and is found in fruits is known as _____fructose_____. Its structure is slightly different from that of glucose and is therefore called an ____isomer_____ of glucose.

6. A carbohydrate made from one type of sugar molecule is known as a ___monosaccharide____ where the prefix __mono___ means one and the root word ___saccharide_____ means sugar.

7. Two examples of a monosaccharides are _____glucose_____ and _____fructose_____.

8. A carbohydrate which consists of two types of sugar molecules is known as a _____disaccharide_____. A common example of this type of sugar is _____sucrose_____ commonly known as table sugar.

9. The two monosaccharides which make sucrose are _____glucose_____ and _____fructose_____.

10 ____Lactose_____ is the carbohydrate found in dairy products. It is classified as a _____disaccharide____ because it is made up of two simpler sugars. Those two are _____glucose_____ and _____galactose_____,

11. The fuel for living things is _____glucose_____.

12. More complex carbohydrates can be broken down into glucose through the action of _____enzymes_____.

    Enzymes primarily work in two ways: _____matchmaking_____ or ____by cutting apart_____.

14. The scientific word for cutting is to _____lyse_____.

15. The enzyme which breaks down table sugar is _____sucrase_____.

16. The enzyme which breaks down milk sugar is _____lactase_____.

17. People who are deficient in producing enough lactase are said to be _____lactose intolerant_____.

18. _____Maltose_____ is a disaccharide found in seeds and grains. The enzyme which breaks down this carbohydrate is _____maltase_____.

19. Plants have the capability of joining several glucose units together into complex structures. These complex structures are known commonly as _____starches_____ and scientifically as _____amylose_____.

20. The enzyme capable of lysing amylose is _____amylase_____ and can be found in our _____saliva_____.

21. _____Cellulose___ is a carbohydrate that is not sweet tasting, but instead functions to provide strength and support to plants. It is formed by very long chains of glucose molecules bonded together.

22. The enzyme which breaks down cellulose is known as ___cellulase_____.

23. Microorganisms within the stomachs of ruminants and termites work to the benefit of their host species. Tell or draw a picture of how this relationship happens.

24. A type of relationship where one organism gains benefit while the other suffers is known as a _____parasitic_____ relationship. Give two examples of this type of relationship.

      1. Answers may vary. Examples: fleas on dog, ticks on dog, flies on sheep, lice on cows, worms inside puppies, etc.

      2.

Name _____ Date_____

Instructions: Below you will find clues to solve this word find puzzle. Refer back to the text portion of the lesson for help. Note that the words may read forward, backward, up, down or at a diagonal.

1. Common name for carbohydrate is a ___sugar__.

2. $C_n(H_2O)_n$ is the generic formula for a ___carbohydrate or sugar_.

3. Carbohydrates are made up of three elements: ___carbon___, _____hydrogen____ and ____oxygen____.

4. Another name for a carbohydrate is a __saccharide____.

5. A carbohydrate made up of only one type of sugar molecule. ____monosaccharide_____

6. A carbohydrate made up of two types of sugar molecules, such as sucrose. ___disaccharide____

7. These compounds can work like matchmakers to speed reactions or scissors to cut apart chemical compounds like carbohydrates. ____enzymes____

8. Scientific name for table sugar is __sucrose____.

9. __Glucose__ is the carbohydrate which fuels living things.

10. This sugar makes the sweetness of fruits. ___fructose____

11. The carbohydrate of dairy products is __lactose_____.

12. Plants store carbohydrates in the form of ____starches_____ and the scientific name is ___amylose_____.

13. This type of carbohydrate is not sweet like other carbohydrates. It forms cell walls in plants and gives plants strength and "crunch." ___cellulose____

14. This enzyme has its action specifically for table sugar. ___sucrose____

15. This enzyme has its action on the sugars found in milk and cheese. ____lactase____

16. People who lack the lactase enzyme are said to suffer from __lactose intolerance_____.

17. __Symbiosis_____ are relationships between living things which may or may not be helpful for each other.

18. In this symbiotic relationship one member benefits while the other suffers. ___parasitism___

19. In this type of symbiotic relationship one member benefits while the other is not directly affected. __commensalism_

20. The enzyme which breaks down cellulose is known as __cellulase___.

21. _Ruminants__ like cows, sheep and deer, have a symbiotic relationship with microorganisms to digest cellulose. Insects, such as _termites_ also have __microorganisms_ in their digestive tracts to digest cellulose.

22. The chemical formula for glucose is __CsixHtwelveOsix__.

SEE WORD COORDINATES AND DIRECTION OF WORD ON NEXT PAGE.

```
     1  2  3  4  5  6  7  8  9 10 11 12 13 14 15 16 17 18 19 20 21 22 23 24 25 26 27 28 29 30 31 32 33 34 35 36 37 38 39 40

 1   E  S  A  T  C  A  L  I  K  O  B  A  N  T  A  P  S  M  Z  A  U  P  V  I  M  T  S  V  X  X  F  U  R  E  I  T  V  Y  K  S
 2   P  M  R  M  V  O  P  H  S  D  N  E  O  K  I  I  W  K  J  E  L  H  Z  T  Q  N  K  Q  B  K  R  F  U  B  C  K  S  U  U  U
 3   D  H  B  C  E  T  D  O  U  F  I  T  K  C  S  U  G  H  L  Z  D  I  K  Z  P  C  C  D  S  F  L  D  M  U  X  P  D  C  R  W
 4   J  H  G  G  Z  I  P  C  F  Q  W  V  N  O  K  W  E  P  B  Q  E  E  Z  G  M  Z  M  C  B  E  C  Z  I  G  W  J  R  N  Y  A
 5   M  S  I  L  A  S  N  E  M  M  O  C  I  E  I  T  D  Q  Y  K  V  N  H  M  R  E  E  U  D  C  X  Z  N  M  Y  A  H  S  Z  S
 6   H  Z  O  U  T  T  Y  E  I  S  Z  B  D  I  P  E  L  A  A  T  L  E  F  J  C  W  P  E  E  N  K  T  A  B  S  R  O  M  H  Q
 7   W  L  Z  Z  X  C  U  J  M  L  M  U  M  A  L  D  Q  D  R  F  O  F  D  X  J  M  P  F  W  A  C  R  N  E  B  Z  U  V  M  Q
 8   H  W  S  U  N  N  B  M  S  Y  J  K  P  U  X  R  O  F  E  T  F  U  W  J  H  C  A  A  T  R  V  K  T  H  X  W  H  C  Q  P
 9   Q  C  E  L  L  U  L  O  S  E  Q  J  N  M  I  L  Y  C  D  L  P  J  A  Z  J  A  Z  R  W  E  Q  P  S  T  G  A  R  E  J  D
10   H  M  D  U  S  B  X  E  T  A  M  W  J  E  A  U  M  D  N  B  L  K  R  Y  W  R  A  F  F  L  Q  D  N  F  L  S  B  A  S  X
11   X  M  R  D  V  E  Y  W  I  I  V  I  A  K  J  N  N  O  S  G  O  J  O  C  X  B  M  H  T  O  D  N  W  U  R  F  D  C  Y  C
12   P  R  K  D  H  I  D  I  Y  N  C  E  C  T  I  C  U  Z  K  G  L  B  K  I  X  O  Y  Y  R  T  G  L  U  C  O  S  E  Q  C  U
13   A  F  Z  S  U  K  V  I  S  U  G  A  R  R  E  H  W  J  V  H  L  Y  S  M  S  H  L  O  A  N  F  R  F  J  N  F  T  T  J  N
14   J  O  O  S  W  X  Z  G  R  Z  S  N  Y  J  O  R  S  Q  W  Y  L  O  F  T  J  Y  O  C  Q  I  M  S  W  E  M  W  C  O  K  G
15   M  O  N  O  S  A  C  C  H  A  R  I  D  E  D  O  S  E  X  H  E  D  V  M  X  D  S  I  B  E  J  C  E  S  O  T  C  U  R  F
16   R  D  T  L  J  I  L  T  O  S  H  W  Y  I  V  R  R  Y  T  V  A  R  B  E  L  R  E  K  J  S  D  G  E  G  D  H  U  X  N  E
17   H  H  G  R  X  I  T  P  C  G  O  C  S  L  A  E  W  G  L  I  P  K  N  N  Q  A  L  G  D  O  U  E  V  L  M  X  X  W  A  K
18   Z  T  A  U  Z  O  M  B  S  I  Q  A  C  N  W  L  O  E  A  Z  M  W  L  C  D  T  Q  W  X  T  W  D  L  Z  L  I  T  R  F  Z
19   B  L  H  U  T  A  U  Z  H  E  C  C  T  A  S  O  W  L  O  N  B  R  G  X  S  E  V  M  H  C  A  D  W  L  S  U  M  K  H  G
20   M  H  I  Z  Y  W  U  C  H  C  G  P  A  O  S  T  M  N  A  A  I  A  E  R  E  T  B  J  P  A  O  B  C  N  U  P  L  Y  Z  J
21   U  L  K  O  H  A  R  O  H  Q  Q  G  I  H  H  E  S  Z  O  C  L  S  F  T  M  J  X  T  N  L  F  F  L  V  M  B  D  A  W  P
22   Z  H  L  I  Q  A  Y  A  N  V  O  E  C  X  S  M  I  K  L  B  T  B  M  M  Y  G  Q  M  L  T  D  E  X  Y  D  R  U  S  S  U
23   A  Y  W  V  T  Q  R  M  E  P  B  A  I  O  D  K  T  J  B  J  R  O  W  S  Z  T  A  K  D  C  O  M  G  E  O  Q  O  A  T  E
24   E  E  X  S  G  I  M  H  G  P  D  S  R  G  Y  A  I  C  T  N  S  A  S  H  N  Z  T  E  N  R  V  V  F  G  W  P  W  H  E  F
25   Q  I  E  Y  D  C  A  D  Y  I  C  C  P  X  R  J  S  V  I  C  F  G  C  E  E  Q  T  W  M  O  Q  N  E  Z  A  H  N  J  C  Y
26   G  H  C  E  X  V  H  B  X  G  U  F  J  Q  P  L  A  A  R  J  J  Q  W  D  K  U  V  G  P  F  Z  N  R  W  S  M  H  W  M  K
27   D  L  M  H  D  A  K  Q  O  S  W  U  P  B  W  F  R  H  K  X  A  D  F  S  F  A  K  T  J  B  H  M  E  T  T  R  M  T  O  B
28   A  I  J  F  B  Q  I  V  V  V  P  W  J  O  P  K  A  O  Y  T  J  P  I  N  V  K  R  L  G  N  C  V  H  H  Q  F  M  Y  E  L
29   O  K  T  Y  U  B  C  Y  N  U  N  L  F  G  O  A  P  N  Q  S  Z  R  H  P  Z  D  Z  V  O  K  B  E  H  J  C  O  T  L  V  U
30   D  J  A  G  W  D  Z  W  R  Y  B  L  S  T  L  F  X  V  H  M  W  U  B  S  U  L  B  V  M  P  G  Q  A  H  L  R  K  X  M  E
```

SOLUTIONS TO WORD-FIND PUZZLE ON PREVIOUS PAGE
Lesson 3, Practice Page 2

Over, Down and Direction

S=south or down, N=north or up, E=east or left, W=west or right.

For example, you would find amylose beginning in the 27th column on row 10 and extending south or down.

AMYLOSE(27,10,S)

CARBOHYDRATE(26,8,S)

CARBON(23,25,NW)

CELLULASE(32,15,SE)

CELLULOSE(2,9,E)

COMMENSALISM(12,5,W)

CSIXHTWELVEOSIX(11,25,NE)

DISACCHARIDE(15,15,SW)

ENZYMES(25,25,N)

FRUCTOSE(40,15,W)

GLUCOSE(31,12,E)

HYDROGEN(39,19,SW)

LACTASE(7,1,W)

LACTOSE(18,19,SE)

LACTOSEINTOLERANCE(30,21,N)

LUNCH(16,9,S)

MICROORGANISMS(11,10,SE)

MONOSACCHARIDE(1,15,E)

OXYGEN(9,27,N)

PARASITISM(17,29,N)

RUMINANTS(33,1,S)

SACCHARIDE(15,20,NW)

STARCH(4,24,NE)

SUCRASE(40,1,SW)

SUCROSE(10,27,NE)

SUGAR(9,13,E)

SYMBIOSIS(9,9,NE)

TERMITES(24,21,NW)

WATER(12,10,SE)

Name _____Date_____

Lesson 4   Practice Page 1

Instructions: Fill-in the blank with the appropriate word. Refer back to the text portion of the lesson for help.

1. Lipids, like carbohydrates, are formed from three elements: _____carbon_____,

___hydrogen__ and _____oxygen_____.

2. There are two main portions to a lipid. The first is a three-carbon component known as - ___glycerol_.

3. The second portion consists of chains of carbons and are known as the _____fatty acid chains__.

4. There can be up to __3_ fatty acid chains per glycerol molecule.

5. A lipid with one fatty acid chain present is known as a _monoglyceride___.

6. A lipid with two fatty acid chains present is known as a __diglyceride___.

7. A lipid with _____3_____ fatty acid chains present is known as a triglyceride.

8. Fatty acid chains can be classified as being __saturated__ or ___unsaturated___ depending upon whether any double bonds are present between the carbon atoms making up the chain.

9. Fatty acid chains which have a double bond present are known as __unsaturated_ fatty acids.

10. Fatty acid chains which have only single carbon bonds present are known as __saturated_ fatty acids.

11. Lipids which are unsaturated usually come from ___plants__ whereas lipids which are saturated come from _____animal___ sources.

12. Unsaturated fats are ___liquid___ in form at room temperature while saturated fats tend to be - ___solid_____ in form at room temperature.

13. Lipids can be converted into fuel (glucose) for living things through the work of ___enzymes___.

14. Because lipids contain so many carbon, hydrogen and oxygen atoms, the relative amount of energy found in one portion of a lipid is about _____9_____ times the amount of energy found in an equal portion of a carbohydrate.

15. Lipids are also found to not mix with water. One can then say that lipids are _____immiscible, insoluble_____ in water or __hydrophobic___.

16. Because of their insolubility in water, lipids are useful for living things when water or watery substances need to be contained in one location or another. (**True** or False)

17. Milk that has all of its fat portion removed is known as ____skim or 0%____ milk.

18. Milk that comes straight from the cow has a fat content of _____3-6%_____.

19. The process whereby the fat portion of the milk (cream) is made to remain equally distributed throughout the watery portion of the milk is known as ___homogenization___.

20. Butter is made from the _____fat or lipid_____ portion of the milk.

21. Explain how butter can be washed with water. <u>Buttermilk is water soluble and will therefore wash away from the butter which is insoluble.</u>

Below are clues to solve the crossword puzzle on the next page.

Across

2. literally meaning afraid of water   hydrophobic

3. form of an unsaturated fat at room temperature   liquid

4. when one substance mixes readily with another   soluble

6. number of bonds carbon atoms desire to form   four

10. long chain of carbon atoms making up portion of a lipid   fattyacidchain

12. when one substance does not mix with another substance   insoluble

13. form of a saturated fat at room temperature   solid

14. source for most saturated fats   animals

15. process whereby lipid portion of milk is equally distributed throughout watery portion of milk homogenization

16. three-carbon portion of a lipid glycerol

17. source for most unsaturated fats plants

Down

1. the quantity of energy a fat has versus the quantity of energy found in an equal portion of carbohydrate   ninetimes

2. literally meaning water loving hydrophilic

4. type of fatty acid when only single carbon bonds are present   saturated

5. one of three elements found in a lipid   oxygen or carbon

7. type of fatty acid when a double carbon bond in present   unsaturated

8. one of three elements found in a lipid oxygen

9. scientific name for a fat lipid

11. one of three elements found in a lipid hydrogen

Lesson 4   Practice Page 2

Instructions: On the previous page, you will find clues to solve this crossword puzzle. Refer back to the text portion of the lesson for help.

For solutions, see previous page!

Name _____Date_____

Instructions: Fill-in the blank with the appropriate word. Refer back to the text portion of the lesson for help.

1. Like carbohydrates and lipids, proteins consist of the elements: ___carbon__, ___hydrogen_ and ___oxygen___. However, proteins also always contain the element __nitrogen_.

2. Proteins consist of smaller components known as ____amino acids___.

3. There are about __50_ amino acids known to man of which there are __20___ which are required by humans to live.

4. Amino acids consist of two portions: the ___common___ portion and then the ____R-group__.

5. The common portion is always the same for each amino acid, but the R-group ___changes_____ which makes amino acids unique.

6. Of the 20 amino acids necessary for humans to live, many can be made from raw materials in our bodies or other amino acids. However, there are __8__ amino acids which cannot be made and therefore must be _____eaten in our diet_____.

7. List those 8 amino essential amino acids: valine, isoleucine, leucine, phenylalanine, threonine, tryptophan, methionine, and lysine.

8. Amino acids are linked to form ___proteins or peptides___. It is the ___sequence_____ of these amino acids which form various proteins needed by the body.

9. The process where amino acids are linked together is known as ___dehydration___ __synthesis___ because water is released through the process.

10. A ___peptide_____ is formed when two amino acids are linking and a __polypeptide___ is formed when multiple amino acids are joined together.

11. ___Enzymes or proteases___ are responsible for dismantling other proteins to provide a source for components to build the amino acids we need. These ___enzymes__ also build these new amino acids.

12. The process of breaking down proteins is known as _____hydrolysis____ which literally means cutting with __water____.

13. Proteins are very large molecules. (**True** or False.)

14. In general, the primary use for proteins in living things is that they serve as __structural or building__ components for the cells of living things.

15. A common protein found in egg whites is __albumin__. This protein is also found in our __blood__ and is responsible for maintaining appropriate amount of ___water___ in our bodies.

16. Another important protein found in our blood is __fibrin__ which is responsible for ___clot_____ formation.

17. And yet another very important group of proteins are the ____immunoglobulins_____ where the prefix immuno– refers to the _immune_ system and globulin refers to __protein__.

18. Immunoglobulins work to fight off _____diseases___ encountered by the living thing.

19. In mammals, the first milk produced by the mother is full of __immunoglobulins_ and has a special name known as _colostrum__.

Name _____Date_____

Lesson 5   Practice Page 2

Instructions: Below you will find clues to solve this word find puzzle. Refer back to the text portion of the lesson for help. Note that the words may read forward, backward, up, down or at a diagonal.

1. Proteins are formed from the elements __carbon__, ___hydrogen__. _oxygen_ and _nitrogen_.

2. Proteins are made of smaller units known as __amino acids___.

3. Amino acids are made with two main components: the ____common portion____ and the ___R group_____ which varies.

4. Amino acids get linked together into peptides through the use of enzymes known as _peptidases__.

5. The bonds which form the linkages between amino acids are known as ____peptide bonds_____.

6. Two linked amino acids are known as a ____dipeptide__ whereas many linked amino acids are known as a ____polypeptide___ where poly– means many.

7. The joining of amino acids whereby water is released Is known as ___dehydration synthesis___.

8. It's the __sequence__ of amino acids which determines which protein is produced.

9. The breaking apart of proteins to create a supply of amino acids is known as ___hydrolysis__ because water is used to cut the protein.

10. The primary use for proteins in living creatures is their role as structural components. A protein found in blood is __fibrin__ aids in blood clotting.

11. Another blood protein is __albumin_ which is the same protein found in egg whites.

12. A final group of proteins are those found in our immune system which are used to fight diseases. This group of proteins are known as the ___immunoglobulins__.

13. Mammals provide a "megadose" of immunoglobulins through the first milk made for their offspring. This first milk is known as ___colostrum_____.

N E N U Y Z E M C P A P X B T V C F J H B D Z P A U O M Y X C T R B R D A G O P
C O O E E S J E E F Y U V C F M A N W K O O E T D F G L M E N J R W E B I G B Q
E J I D G R O P V Y V S X L M M R U L S Q Y I K P A N S C P W P H H Z E R E Z J
P B L T P O T M U R T S O L O C B E O P T F J W B F X X D R T W Y C L L W E J P
U F C I R I R I T G Q E F Z I Q O I J F I W L E C O H A P I N D E N Z Y M E S C
L S T M D O O D G Q Q J N K F Y N K O E Z T B C G E Q G F C R O N P O T P R S L
K Y C A V X P V Y B W F R I X N C W Z M V X D K K P H J A A G Y V O U N I R Y U
D H S N Y D F N F H O U A L R D I S T D B T G B C J L Q T L S G D I Y E J L P U
B E Y G Q E W A O F D W M U E B B K S M A M I N O A C I D S B Y I H X G M N P S
E L E V U N W W I M P U O R G R I N O W D T Z M Z K O P Z O S U X V A O M I M Q
E N R H W X S X J W M W Q P Y M K F N R B Q L J T N M I L I M G M D F R G Z I V
K C S A Z Z F D Y N Q O X U R L X S K B A V X C S I S D E E B J Q I O T X G V O
R C J H U B V Y M B U F C S U S P M R L O M N Y X J N D Z O X I O L N I M C B Z
G A X U B C O P R N P Q B A F Q W I P X C T N C M J E F X Y K C X H K N A P A D
M S Q L C G B D K F H W M N Z J O T Z J S T B A O K H R A B O M Y P I G R O J R
Z E G D O W Z N S V A T S X V J Z X O J H Z L X T Q I X X V T S U A J T C U Y K
F I A O X I B O M L Z N F D L O W X E E S M A D Y U O K U O N Z B A O X M N X N
U N N C W U K B T L V G I Q S K V G S T A S N I L U B O L G O N U M M I B T A H
H I Y P G Q T E I Q W E W Y O O R I X O C S D E Q Z X O S M L C T M F N Q A P R
D Y J K U C G D L Y M B I L E N S O K U Y F F Y K W C S K K B D D F X E U E F V
I A D D U A Y I D U G V W D M T Y I E S H V G Z D E Q Y M C J R H Q Z R T N V V
P J S R M T M T A X O S I H F O L L H F D O E R M S O J E K H E Y M V N H Q X T
E B U C O N H P B W U T D E V W D G I F W Z M X N D A Q M U S T X A M A C L O X
P C O X Y L D E L R P O C P G Q R O Q I P A W Y Y W Y Z B V G A Q Q T W Z L C Y
T S L Q P A Y P F E S N P R J E V N L C O B C P D J F P E C Q W L M D A B P A U
I Y R W N E B S P Q E N O Y I T F K V R J L Q V H Q D G R R U K B L B C P A B C
D B O U I H I Y I U N B J K M U S G R N X P Z O A V T D D I D O B C I P K Y K Y
E W K X D P L K Q S I P E Y E A M X Z T Z K T Y H E O I X I N S F E V N V O G P
Y J Q I H O Z E Y X A N J Z B U P L O Y O N Q B J B N C H E R G H B K D R B O U
U W P H P L S K W J B I Z M L I V S Q I J S B D A V X G B R S Y O R M P F E N E

Name _____Date_____

Lesson 6   Practice Page 1

Instructions: Fill-in the blank with the appropriate word. Refer back to the text portion of the lesson for help.

1. All of the uses for carbohydrates, proteins and lipids in living things, whether the building or breaking down of these molecules are dependent upon the presence of __enzymes___.

2. And, these enzymes are very dependent upon the __pH____ of the environment in which they are working.

3. With regard to the symbol used for pH, the lower case p is thought to represent the term __percent___ while the H stands for _____hydrogen____.

4. Substances which have a desire to get rid of hydrogen ions are said to be _acids__.

5. Substances which have a desire to accept hydrogen ions are said to be __bases__.

6. The pH scale tells us the relative strength of acids and bases. As one moves from value to value on the pH scale the relative strength is multiplied ___10___ times and the scale is referred to as being a __logarithmic___ scale.

7. So an acid with a pH of 3 would be _____100_____ times stronger than an acid with a pH of 5.

8. On the pH scale, acids are less than _____7_____ while bases are greater than _____7_____. Bases taste __bitter___ while acids taste _____sour_____.

9. Secretions in our stomachs are considered strong acids. What value on the pH scale would you expect if you measured the pH of these gastric secretions? _____0-1_____

10. Vinegar or _acetic acid_ is a ____weak___ acid but still is effective at preserving foods. This is because the __bacteria causing spoilage cannon survive at this low of a pH_____.

11. Tell the pH of the following body substances: Blood _____7.4_____ Milk ___6.3-6.6___ Pancreatic secretions ___8.1___ Cerebrospinal fluid ___7.5____ Urine ___6.0_____

12. By raising or ____lowering_____ the pH of fluids of living things we can directly affect the enzymes functioning in that living thing.

13. pH can be measured by colored strips of paper which have a powder affixed to them. These strips of paper are known as ___litmus paper__.

14. Red litmus paper will turn __blue__ in the presence of a base.

15. Blue litmus paper will turn ___red___ in the presence of an acid.

Name _____Date_____

Lesson 6   Practice Page 2

Instructions: Below you will find clues to solve this crossword puzzle.

Across

4. base used to leaven cookies
baking soda

6. greater than seven on the pH
scale base

8. acids taste sour

9. bases taste bitter

11. the pH scale's difference between values logarithmic

13. pH of blood sevenpointfour

14. color of litmus paper in presence of acid red

Down

1. scientific name for vinegar
acetic acid

2. what the H stands for in pH
percent

3. pH of urine six

4. color of litmus paper in presence of base blue

5. pH strips capable of testing
range of pH values hydrion paper

7. less than seven on pH scale
acid

10. enzymes are dependent upon
__pH___ to function correctly

12. what the p stands for in pH
percent

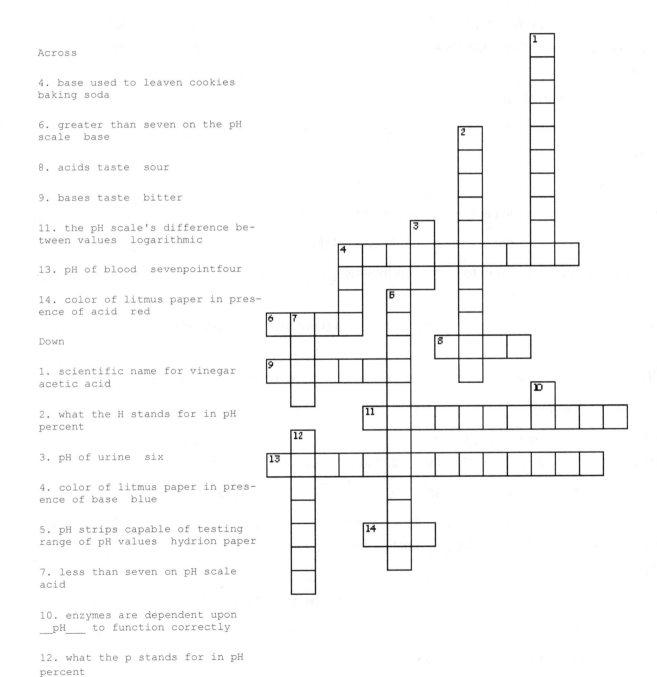

Name _____Date_____

Lesson 7   Practice Page 1

Instructions: Fill-in the blank with the appropriate word. Refer back to the text portion of the lesson for help.

1. The study of cells is known as _____cytology_____.

2. The scientist who first made the discovery of cells was _Antonie van Leeuwenhoek _ in __1600's_____. He gave cells their name because it is said he though they resembled _____jail cells_____.

3. The "little organs" within cells are known as _____organelles_____.

4. Cells are covered with a "skin" known as the ____cell membrane_____ or ___plasma membrane_____.

5. The cell membrane consists of a ___phospholipid___ bilayer.

6. The phosphate groups are _____hydrophilic___ while the lipid layer is _____hydrophobic_____.

7. This allows the membrane to be able to control the passage of __substances__ .

8. Cell membranes are said to __semipermeable___ meaning that the cell has the ability to allow certain substances to pass but not others.

9. Plants have an extra layer of protection around their cells. This layer is known as the ___cell wall____.

10. Cell walls are made of ___cellulose___ which are long strings of __glucose__ molecules.

11. Cells walls function like __a skeleton__ for plant cells giving them strength and support.

12. Trees which have thin cell walls and tend to grow quickly are known as ____softwoods___ while trees which have thick cell walls and tend to grow slowly are known as _____hardwoods_____.

13. The cell organelle which functions like the brain of the cell is the ___nucleus___.

14. The nucleus, depending upon which creature it is in, may or may not have a covering membrane. This membrane is known as the ____nuclear membrane_____ or ____nuclear envelope____.

15. Organisms which DO have a nuclear membrane are identified as being ___eukaryoytes_____ .

16. Organisms which do NOT have a nuclear membrane are identified as being _prokaryoytes____.

17. An example of an organism which is a prokaryote is a __bacteria_____.

18. The structures within the nucleus of a cell which hold the instructions for the cell are the

___chromosomes____.

19. Organisms are specific for the __number__ of chromosomes they have and for the information found on each chromosome.

20. Humans have __23_____ pairs of chromosomes for a total of ___46___ chromosomes in most of their cells.

21. Within chromosomes we find subunits or "recipes" which are known as ____genes_____.

22. Genes code for __features or traits or characteristics____ in the organism.

23. The organelle of cells which functions like a power plant is the __mitochondrion__.

24. The fuel utilized by the power plants of cells is ____glucose____.

25. The chemical formula for glucose is ____$C_6H_{12}O_6$_____.

26. Mitochondria are shaped like ___beans_____.

27. The inner membrane of a mitochondrion is known as _____cristae____ and this is where ____respiration__ takes place.

28. Respiration is the process whereby glucose is converted into __energy_ which is in the form of ___ATP__s.

29. ATP is the abbreviation for _adenosine triphosphate___.

30. In the presence of oxygen, one glucose molecule can yield _____38_____ ATPs.

31. This form of respiration is known as _____aerobic____ respiration.

32. Where there is little or no oxygen present, one glucose molecule can only yield ___2_____ ATPs.

33. Respiration with little to no oxygen present is known as _____anaerobic___ respiration.

34. The two main uses of ATPs are to allow for _____movement____ and __heat_ production.

Lesson 7   Practice Page 2

Instructions: On the next page, you will find clues to solve this crossword puzzle. Refer back to the text portion of the lesson for help.

Clues for Crossword Puzzle

Across

4. loves water   hydrophilic

5. fuel used by mitochondrion of cell   glucose

6. what discover of cells thought they looked like   jailcells

11. what ATP stands for   adenosinetriphosphate

15. hates water   hydrophobic

16. total number of chromosomes in most of a human's cells   fortysix

17. aerobic respiration makes this many ATPs.   thirtyeight

19. "skin" of cells   cellmembrane

21. anaerobic respiration results in this many ATPs.   two

22. process whereby glucose is converted into ATPs.   respiration

23. chemical formula for glucose   CsixHtwelveOsix

Down

2. scientific term which describes how cell membranes are composed of two layers of phosphates and fat compounds   phospholipidbilayer

3. "power plant" of cell   mitochondrion

7. thin cell wall of pine or fir   softwood

8. skeleton of plant cell   cellwall

9. study of cells   cytology

10. carbohydrate which makes cell wall   cellulose

12. means allows some things to pass but not others   semipermeable

13. colored bodies which are like chapters in the cell's cookbook   chromosomes

14. "recipes" found in the chromosomes   genes

18. thick cell wall of oak or maple   hardwood

20. "brain" of cell   nucleus

Name _____Date_____

Lesson 8   Practice Page 1

Instructions: Fill-in the blank with the appropriate word. Refer back to the text portion of the lesson for help.

1. The organelle of cells which is responsible for packaging of products for export from the cell is the _____golgi body_____.

2. When a portion of golgi body is filled with product and ready to be shipped from the cell, a portion known as a ____vesicle____ breaks free to moves toward the periphery of the cell.

3. The organelle of a cell which functions primarily as storage or, in some cases, locomotion, is the _____vacuole_____.

4. The part of a cell which functions like a transportation system is the __endoplasmic__ ____reticulum___ or ___ER___ for short.

5. There are two types of ER: ____smooth____ and ___rough____ ER.

6. The roughness is rough ER is due to the _____ribosomes_____ present.

7. The function of ribosomes are to build _____proteins_____ from available __amino acids____.

8. Another role performed by ER is the folding of _____proteins_____.

9. There are four degrees of protein folding: __primary___, ___secondary____, ____tertiary_____ and _____quaternary____.

10. The functionality of a protein is dependent upon the way it is ___folded____.

11. Sarcoplasmic reticulum is found in _____muscle____ cells and is important in the regulation of _____calcium_____.

12. _____Microtubules_____ are like a microscopic skeletal system for cells. They can enable movement, too.

13. Microtubules which make up centrioles are important in ___mitosis/cell division_____.

14. _____Chloroplasts_____ are organelles found in green plants which are capable of the process of _____photosynthesis_____.

15. The raw materials for photosynthesis include: ___carbon dioxide_, ___water____, and energy from the _____sun_____.

16. The product of photosynthesis is _____glucose_____ which has the chemical formula of _____$C_6H_{12}O_6$_____.

17. This glucose can be used by _____plant_____ or can be "stolen" by another living creature to be used as a _____source of energy_____.

18. A waste product from cellular ___respiration__ is ____carbon dioxide___ which becomes a resource for plant use in photosynthesis.

19. Plants and creatures which undergo cellular respiration are ___dependent__ upon each other.

Clues and Answers for crossword puzzle on next page.

Across

6. fourth degree of protein folding   quaternary

9. transportation system of cell endoplasmic reticulum

12. second degree of protein folding secondary

13. ER with ribosomes   rough

17. formula for glucose   CsixHtwelveOsix

20. little protein building factories found on rough ER   ribosomes

Down

1. photosynthesis factory of plants   chloroplasts

2. reticulum of muscles   sarcoplasmic

3. closet for cell   vacuole

4. process of making glucose from carbon dioxide, water and sunlight   photosynthesis

5. this element is controlled by sarcoplasmic reticulum   calcium

7. third degree of protein folding tertiary

8. we are _____ upon plants and they are _____ upon creatures like us   dependent

10. works like a UPS man   golgibody

11. tiny bones of cells   microtubules

14. first degree of protein folding primary

15. byproduct of photosynthesis but useful for animals   oxygen

16. important in cell division   centrioles

18. ER without ribosomes   smooth

19. package ready for shipment from cell vesicle

Lesson 8   Practice Page 2

Instructions: Below, you will find clues to solve this crossword puzzle. Find clues for this puzzle on the previous page.

ANSWERS ON PREVIOUS
PAGE

Name _____Date_____

Lesson 9   Practice Page 1

Instructions: Fill-in the blank with the appropriate word. Refer back to the text portion of the lesson for help.

1. Living things which are made up of many cells grow in size by ___increasing___ the number of cells from which they are made.

2. The process of cell division is known as ____mitosis_____.

3. The phase of a cell's life cycle when it is going about its assigned job is known as ____interphase_____.

4. In animals and humans stimulation to divide usually comes in the form of a _____hormone_____.

5. Hormones are substances usually made by ___glands____ in one part of the body and have their effects in other parts of the ____body_____ being carried there by the _____blood_____.

6. Growth hormone is made by the ____pituitary gland_____ which is located _____beneath the brain_____.

7. The cell that receives the signal to divide is referred to as the ____parent__ cell and the two cells which result from the division are known as the _____daughter_____ cells.

8. The first stage of mitosis is known as ____prophase_____.

9. In prophase the ___chromatin__ which was originally uncoiled while being used by the cell, ____contracts___ or bunches up and becomes visible again as chromosomes.

10. The _____nuclear membrane_____ also begins to break apart during this stage.

11. The second main event of prophase is that the chromosomes of the parent cell must _____make copies of themselves/duplicate_____.

12. When copying of the chromosomes does not happen correctly, a ___mutation_____ is said to have occurred.

13. The second stage of mitosis is known as ____metaphase_____. During this phase the chromo-somes align themselves along a _central plate_ within the cytoplasm of the cell. ___Spindle___ fibers attach themselves to the chromosomes.

14. The next stage of mitosis is known as ____anaphase_____.

15. During anaphase, the spindle fibers ____contract_____ pulling their attached ___chromosomes_ to opposite poles of the cell.

16. The final stage of mitosis is known as _____telophase_____.

17. During telophase the chromosomes begin to _____uncoil_____ and the cell membrane begins to _____pinch inward____ eventually allowing for the cell to __divide/split__ into __two daughter__ cells.

18. The process whereby the cell membrane presses inward to create two daughter cells is known as _____cytokinesis_____.

19. The two "new" daughter cells return back to ____interphase___ where they continue their jobs.

Solutions to Word Find Practice Page 2

(Over, Down, Direction)

ANAPHASE (12, 1, SW)

CENTROMERE (5, 12, NE)

CHROMATIN (7, 12, E)

COILUP (5, 6, NE)

COPIES (2, 10, N)

CYTOKINESIS (13, 15, W)

DAUGHTERCELLS (13, 2, SW)

INTERPHASE (15, 10, N)

METAPHASE (5, 3, SE)

MITOSIS (13, 7, NW)

PARENTCELLS (15, 11, NW)

PLATE (10, 13, E)

POLES (9, 1, SE)

PROPHASE (1, 1, S)

SPINDLEFIBERS (4, 1, S)

Name _____Date_____

Lesson 9   Practice Page 2

Instructions: Below you will find clues to solve this word find puzzle. Refer back to the text portion of the lesson for help. Note that the words may read forward, backward, up, down or at a diagonal.

Phase of mitosis where chromosomes are moving to opposite poles of cell. anaphase

Part of chromosome where spindle fiber attaches. centromere

When chromosome is uncoiled and being used by the cell, this genetic material is called ___chromatin___.

During prophase, the uncoiled chromatin will __coil up_ to form chromosomes.

____Copies_____ of the chromosomes are made during prophase.

During telophase, the cell membranes pinches inward, a process known as _cytokinesis_.

Parent cells divide into two _daughter cells_.

Phase of cell life cycle where "regular job" is taking place. interphase

Stage of mitosis where chromosomes line up along central plate or equator of cell. metaphase

Cell division = __mitosis____

__Parent cells____ divide to form daughter cells.

During metaphase, the pairs of chromosomes line up along a central ___plate___ or equator of the cell.

During anaphase the chromosomes get pulled to opposite __poles__ of the cell.

During ___prophase_, chromatin coils up to form chromosomes and the chromosomes duplicate themselves.

____Spindle fibers__ attach to chromosomes during metaphase to eventually pull the chromosomes to opposite poles of cell.

ANSWERS ON PREVIOUS PAGE

```
P N F S S B S L P P V A T O E
R E H P P L G I U O N N D H S
O R C I M V L L S A L A Y E A
P G X N C E I E P O U E R E H
H S N D P O T H C G T E S W P
A E P L C J A A H T M I A X R
S I X E L S W T P O N K M O E
E P I F E X E K R H L E E J T
M O A I G R T T N S A H R S N
W C J B C L N F V E X S C A I
H T V E T E E S A H P L E T P
L J L R C C H R O M A T I N
Y L C S Q S G C G P L A T E H
S H T E X N A V Z G H W P S D
S B S I S E N I K O T Y C R T
```

101

Name _____Date_____

Lesson 10   Practice Page 1

Instructions: Fill-in the blank with the appropriate word. Refer back to the text portion of the lesson for help.

1. Chromosomes must _____duplicate__ themselves in order for each daughter cell to have its own set of chromosomes.

2. Duplication of chromosomes is made possible through the function of ___enzymes____ which work to either ___lyse___ or "matchmake" the appropriate parts of the DNA.

3. DNA is the abbreviation for ____deoxyribonucleic acid_____.

4. The carbohydrate component of DNA is a ___ribose____ sugar.

5. The nucleotide portion consists of the _____ribose_____ sugar, a ____phosphate_____ group and then a pair of ____bases_____.

6. The "rung" portion of the DNA ladder is made up of pairs of ____bases_____.

7. There are two groups of bases: the _____purines_____ and the _____pyrimidines_____.

8. The two purines are _____adenine_____ and _____guanine_____.

9. The two pyrimidines are _____cytosine_____ and _____thymine_____.

10. _____Adenine___ always bonds with _____thymine_____ while ____cytosine_____ always bonds with ___guanine_____.

11. The bonds between the bases are weak bonds known as _____hydrogen___ bonds.

12. Enzymes ____lyse/cut_____ the hydrogen bonds exposing the base pairs.

13. Matchmaking enzymes bring in ___free-floating nucleotides_____ in the area joining them according to the matching rules.

14. Eventually, ____two____ new ____strands___ of DNA are formed; one strand for each _____daughter_____ cell.

Name _____Date_____

Lesson 10   Practice Page 2

Instructions: On the next page, you will find clues to solve this crossword puzzle. Refer back to the text portion
of the lesson for help.   SOLUTIONS FOUND ON NEXT PAGE.

SOLUTIONS TO CROSSWORD PUZZLE ON PREVIOUS PAGE
Lesson 10, Practice Page 2

Across

1. always bonds with guanine   cytosine

3. group of bases containing adenine and guanine   purines

5. always bonds with thymine   adenine

6. "new" cells which each need their own copy of the DNA cookbook.   daughter cells

7. Chromosomes must be __duplicated_____ so that each daughter cell gets a complete copy of all DNA.

8. These are the recipes within each chapter of the genetic cookbook.   genes

10. The sugar found in DNA.   ribose

11. The stronger bond between the ribose sugar and phosphate group of a nucleotide is a _covalent_____ bond.

12. group of bases containing cytosine and thymine   pyrimidine

13. These are the chapters within the genetic cookbook. chromosomes

14. Part of nucleotide with phosphorus present.   phosphate group

15. these unzip DNA to expose the base pairs   enzymes

16. Parts of nucleotide.   bases

17. Abbreviation for deoxyribonucleic acid.   DNA

Down

2. Shape of DNA discovered by Watson and Crick.   Twisted ladder

4. Type of bond found between base pairs that is unzipped by enzymes.   hydrogen

9. Chromosomes are made of long strands of _deoxyribonucleic acid____ .

Name _____Date_____

Lesson 11   Practice Page 1

Instructions: Fill-in the blank with the appropriate word. Refer back to the text portion of the lesson for help.

1. The making of proteins is known as protein _____synthesis_____.

2. The "recipe" for making a protein correctly is found in one's __DNA/genes__.

3. There are four nitrogen bases found in DNA and they are ____adenine_____, ____thymine_____, ____cytosine_____ and ____guanine_____.

4. And, it is the _____sequence_____ of these bases which determines the type of protein that can be synthesized.

5. Before a protein can be made, the particular section of DNA must be ___unzipped/opened___ by ____enzymes_____ to expose the base pairs.

6. After the base pairs are exposed, __RNA____ can be formed to make a "copy" of the sequence of the bases.

7. Unlike DNA, RNA is ___single___ stranded and contains the base ___uracil____ instead of ____thymine___.

8. The RNA which forms on the exposed DNA bases takes the name ____messenger____ or mRNA for short.

9. The formation of mRNA takes place in the _____nucleus_____ of the cell whereas protein synthesis takes place out on/in the _____cytoplasm_____ of the cell.

10. More specifically, proteins get built in the _____ribosomes_____ found on the ____rough____ endoplasmic reticulum

11. Out at the ribosome, available amino acids are "tagged" with ___transfer RNA__ or tRNA for short.

12. The tRNA is specific for each _____amino acid_____ and will therefore only fit the appropriate sequence of the mRNA that has arrived at the ribosome. The tRNA works like a ____key____ that fits the mRNA "lock."

13. The tRNA is in sets of _____3_____ base pairs and is called an __anticodon___.

14. Specific codons code for specific _____amino_____ acids.

15. As the amino acids get aligned to the mRNA, _____enzymes (got to love those enzymes)_____ work to link together forming _____peptide_____ bonds. Long strands of peptides form in this way.

16. The process whereby mRNA is made from a section of DNA is called _____transcription_____ where the process where tRNA assembles amino acids in the ribosome is known as _____translation_____.

17. When one's DNA is altered or damaged, the proteins made from this damaged DNA may not _____function/work_____ correctly. A change in the DNA is called a _____genetic_____ mutation.

18. Things that cause mutations are called ___mutagens_____ or ___teratogens___.

19. An example of a mutagen is _____radiation____ which is energy that can come from the sun or an elemental source.

20. Cells which have had their DNA altered may become ____cancer___ cells and change their behavior.

21. Cancer treatments may include bombarding cancer cells with radiation in an attempt to cause ____mutations___ in the DNA of those cells leading to their destruction.

Name _____Date_____

Lesson 11   Practice Page 2

Instructions: Below are 17 words or phrases that we've introduced in this lesson. Their spellings have been scrambled. Rearrange the letters to spell these words and place them into the boxes. Transfer the letters in the numbered boxes to the "finale" at the bottom. If you need hints on unscrambling the words, turn the page.

QESNECUE   □□□□□□□□ (16)

MGNERSERNASE   □□□□□□□□□□□□ (18, 24, 25)

SENFARANTRR   □□□□□□□□□□□ (5, 21)

NAD   □□□ (19)

BOIRESMOS   □□□□□□□□ (23, 3, 20)

LUUSECN   □□□□□□□ (8)

MIENIRASOLLUCTEDUCPM □□□□□□□□□□□□□□□□□□□□□ (1, 17)

GEURHOR   □□□□□□□ (2, 12)

ZEYNMES   □□□□□□□ (22, 9)

DOPSBDETPENI   □□□□□□□□□□□□ (4)

MANODSICIA   □□□□□□□□□□ (6)

SONCDO   □□□□□□ (10)

RANSOITLNAT   □□□□□□□□□□□ (7)

SRNAITOINTRPC   □□□□□□□□□□□□□ (11)

TOTMANSIU   □□□□□□□□□ (14)

ANRIATIOD   □□□□□□□□□ (15)

REACNC   □□□□□□ (13)

Solutions:

sequence

messengerrna

transferrna

dna

ribosomes

nucleus

endoplasmicreticu-
lum

rougher

enzymes

peptidebonds

aminoacids

codons

translation

transcription

mutations

radiation

cancer

□□□□□□□ □□□□□□□□□ □□ □□□□□□□.
1 2 3 4 5 6 7 8 9 10 11 12 13 14 15 16 17 18 19 20 21 22 23 24 25

107

Hints for solving the scrambled words puzzle are found below. Note that these hints are not in the same order as the puzzle words.

Transcribing molecules

These get linked together to form proteins.

It is the _____ of bases which determines the protein synthesized.

These make smooth ER not smooth.

Sets of threes.

Location of DNA

ER

ER that is not smooth.

Tags or keys found on amino acids

Scissors or matchmakers

Bonds between amino acids.

When genes gets messed up.

Rays from the sun.

Disease where cells are out of control.

Ladder-like molecule with all information.

mRNA does this job

tRNA does this job

Name _____Date_____

Lesson 12   Practice Page 1

Instructions: Fill-in the blank with the appropriate word. Refer back to the text portion of the lesson for help.

1. The process by which living organisms create more complete living organisms is known as _____reproduction_____.

2. Organisms with two parents have pairs of _____chromosomes___, one member of each pair coming from each ____parent____.

3. Humans have ___23___ pairs of chromosomes for a total of ___46_____.

4. The variation in a trait is known as an _____allele_____. Examples of alleles for eye color could be: ___blue, brown, hazel_____.

5. When it comes to alleles, chocolate is to ice cream as __red, brown, black, brunette, blond, etc.____ is to the gene for hair color.

6. Living creatures which have pairs of chromosomes are said to be __diploid__ or 2N.

7. Almost all cells in humans and animals are 2N except for the __sex__ cells or _____gametes_____.

8. These cells (sex cells or gametes) are __1__N or haploid.

9. The cells which eventually become gametes are known as the __primordial___

_____sex cells_____ and originally are __2___N.

10. The primordial sex cells undergo the process of ____meiosis___ to reduce the number of chromosomes from 2N to __1N_____.

11. In human males, one primordial sex cell will result in _____4_____ sperm cells.

12. In human females, one primordial sex cell will result in _____1_____ ovum and three cells which do __not____ survive.

13. Human males continue to create more primordial sex cells throughout their lives, while human females are born with the total number of primordial sex cells they will have their entire lives. (**True** or False)

14. In males, the process of creating sperm cells takes place in the ___testis/testicles_____.

15. In females the process of creating ova takes place in the _____ovaries_____.

16. The joining of the 1N sperm cell with the 1N ovum to create a 2N individual is known as

_____fertilization_____.

17. Reproduction which involves two parents, a male and female each contributing a single cell for the new offspring, is known as _____sexual_____ reproduction.

18. Reproduction which involves only one parent organism is known as ___asexual___ reproduction.

19. In _____sexual_____ reproduction there is a great possibility for an assortment or mixing of genetic information from each parent.

20. In _____asexual_____ reproduction, the offspring carries the exact same genetic information as the parent. There is not any change.

21. Bacteria utilize a means of asexual reproduction whereby the "parent" bacteria creates a copy of its genetic material and then promptly splits into two equal halves, each with its own copy of genes. This method of reproduction is known as ___binary fission___.

22. Another means of asexual reproduction is where genetic material is isolated near the outside of the organism. The cell membrane encircles the genetic material and other cellular organelles forming a bud. The bud eventually breaks free. This method is known as _external budding___.

23. Some plants, like strawberries and spider or airplane plants create "baby" plants by sending out special stems called _____stolons_____ which eventually touch ground forming roots and new leaves. This method that plants use is called __vegetative___ propagation___.

24. The plants which result from vegetative propagation will have genetic information unlike the parent plant. (True or **False**)

25. Ferns and mushrooms reproduce by producing spores. This method of reproduction is known as ___sporogenesis___.

26. Some creatures, like starfish for example, are able to regenerate whole new living organisms from broken parts. This form of reproduction is known as __fragmentation__.

27. Two parent involvement = _____sexual_____ reproduction.

28. One parent involvement = _____asexual_____ reproduction.

29. Sexual reproduction = mixing of genetic information in offspring. (**True** or False)

30. Asexual reproduction = no mixing of genetic information in offspring. (**True** or False)

Name _____Date_____

Lesson 12   Practice Page 2

Instructions: On the next page, you will find clues to solve this crossword puzzle. Refer back to the text portion of the lesson for help.   SOLUTIONS FOUND ON NEXT PAGE.

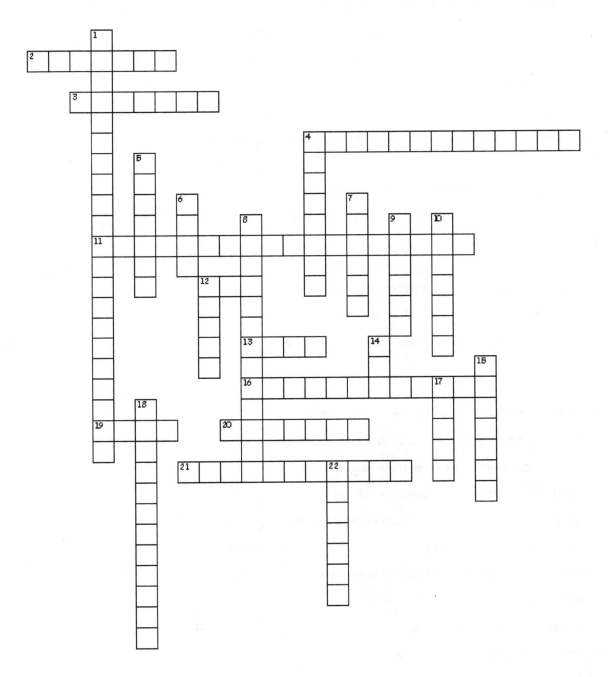

SOLUTIONS TO CROSSWORD PUZZLE ON PREVIOUS PAGE
Lesson 12, Practice Page 2

Across

2. term used for sex cells   gametes

3. process where a diploid cell becomes haploid   meiosis

4. joining of ovum and sperm to create new individual   fertilization

11. cells which are 2N and destined to undergo meiosis to become 1N   primordial sex cells

12. female gametes   ova

13. number of gametes created in males for each primordial sex cell 1   four

16. process of reproducing through the formation of spores   sporogenesis

19. haploid   oneN

20. having pairs of chromosomes   diploid

21. number of pairs of chromosomes in humans   twentythree

Down

1. means of reproducing often employed by plants where specialized stems or roots grow away from parent plant with small "baby" plant forming on end of stem   vegetativepropagation

4. total number of chromosomes in humans   fortysix

5. asexual form of reproduction where a small portion of parent cell emerges and then breaks free from parent cell   budding

6. diploid   twoN

7. reproduction where two parents are involved   sexual

8. form of reproduction where parent cell divides into two new cells   binary fission

9. location in males where gametes are produced   testis

10. variations of a trait on a chromosome   alleles

12. location in females where gametes are produced   ovary

14. number of gametes created in females for each primordial sex cell   one

15. reproduction where only one parent creates new living organism   asexual

17. male gametes   sperm

18. process where a living organism creates more living organisms like itself   reproduction

22. having half the number of chromosomes   haploid

Name _____Date_____

Lesson 13   Practice Page 1

Instructions: Fill-in the blank with the appropriate word. Refer back to the text portion of the lesson for help.

1. The biologist who discovered many concepts of genetics was named _Gregor Mendel___.

2. One of Mendel's biggest contributions was that traits are __inherited from one's parents_____.

3. The record of genes found in a living organism is referred to as the __genotype_____.

4. The results of the genotype that can be observed whether physical attributes or behaviors is referred to as the _____phenotype_____.

5. The possible variations of a gene are referred to as _____alleles_____.

6. We can think of alleles as the possible __flavors____ available for ice cream or cookies.

7. The Law of _____Segregation_____ says that, in meiosis, one chromosome from each parent primordial sex cell will go to each resulting gamete.

8. The Law of ___Independent Assortment___ says that traits get distributed independently from other.

9. When looking at the genotype of the offspring that was created through sexual reproduction, we find ___two alleles_____ present, one from the male parent and one from the female parent.

10. While both alleles are present, only one gets expressed. The __dominant___ allele, if present, is one that always gets expressed.

11. The allele that is present, but does not get expressed is the _____recessive_____ allele.

12. When describing genes, one usually uses __letters of the alphabet____. To indicate dominant alleles, one uses ___upper____ case letters and recessive alleles are indicated by ____lower_____ case letters.

13. When dominant alleles are inherited from both parents, for example HH, the individual is said to be ___homozygous___ dominant for the H allele.

14. When recessive alleles are inherited from both parents, for example hh, the individual is said to be __homozygous recessive____ for the H allele.

15. When an individual inherits a combination of a dominant and recessive allele for a specific trait, that individual is said to be _____heterozygous_____ for that trait.

16. Suppose hair color is indicated using the letter "H." Suppose, also, that black hair color is dominant and red hair color is recessive. If this is the case, what color hair would these individuals have: HH _____black_____ Hh_____black_____ hh___red____

17. If Tom has black hair (from question 16 above,) what are his possible genotypes? _____HH_____ or _____Hh_____

18. If Tom has red hair, what is his only possible genotype? _____hh_____

19. If Tom is heterozygous for black hair color, what must his genotype be? _____Hh_____

20. If Tom is homozygous dominant for black hair color, what must his genotype be? __HH__

21. If all of Tom's biological children have black hair, regardless of the hair color of his wife, what must Tom's genotype be? _____HH_____

22. If Tom has some children with black hair and some with red hair, what must his genotype be? _____Hh_____

23. If Tom has the genotype Hh and his wife has the genotype Hh, do they have the chance of having a child with red hair? _____Yes_____

24. If Tom has red hair and he marries his wife who has black hair, could they ever have children with red hair? _____Yes____ How can this be? _____His wife could be Hh_____

25. If Tom has red hair and he marries his wife who also has red hair, will they ever have children with black hair (according to our "story" of hair color dominance)? ___No____

26. If Tom has black hair and his wife has black hair, could they ever have the chance of having a child with red hair? _____Yes_____ If so, how could this occur? __If they were both heterozygous (Hh) _____.

27. Suppose Tom is homozygous dominant for black hair color and his wife is homozygous dominant for black hair color. Will they ever have the chance for having children with red hair? _____No_____ Why or why not? _____No h allele is present._____.

28. The situation where the phenotype of an individual appears to be the result of sharing of two dominant alleles is known as _____co-dominance or incomplete dominance___.

29. When one finds that certain traits, such as down color of chicks, is found only in one gender, the trait is said to be _____sex-linked_____.

Name _____Date_____

    Suppose you have some show quality cats that you would like to breed to be able to sell their kittens.   While all of the kittens bring good profits, the ones with long hair bring the most returns. Suppose that hair length in cats utilizes the letters "L" and "l" for the genotype. Suppose, also, that long hair length is a recessive trait.   Look at the following situations and predict the possible out-comes for hair length in your kittens. Use the Punnet squares to support your predictions. Male cats are referred to has toms and females are referred to as queens.

A. Your tom is homozygous recessive (l l) and your queen is heterozygous (Ll) for hair length.

	l	l
L	lL	lL
l	ll	ll

Number of kittens with

Long hair: _____2_____

Short hair: _____2_____ out of every four kittens.

B. Your tom is heterozygous (Ll) and your queen is homozygous (l l) recessive for hair length.

	L	l
l	Ll	ll
l	Ll	ll

Number of kittens with

Long hair: _____2_____

Short hair: _____2_____ out of every four kittens.

C. Your tom is heterozygous and your queen is heterozygous for hair length.

	L	l
L	LL	lL
l	Ll	ll

Number of kittens with

Long hair: _____1_____

Short hair: _____3_____ out of every four kittens.

D. Your tom is homozygous recessive and your queen is homozygous recessive.

	l	l
l	ll	ll
l	ll	ll

Number of kittens with

Long hair: _____4_____

Short hair: _____0_____ out of every four kittens.

E. Your tom is homozygous dominant and your queen is homozygous dominant.

	L	L
L	LL	LL
L	LL	LL

Number of kittens with

Long hair: _____0_____

Short hair: _____4_____ out of every four kittens.

Which situation (question letter, A-E) would potentially be the best money-making opportunity for you? _____D_____

Name _____Date_____

Lesson 14   Practice Page 1

Instructions: Fill-in the blank with the appropriate word. Refer back to the text portion of the lesson for help.

1. The study of the form of living things is known as _____morphology_____.

2. The study of grouping organisms based upon similarities and differences is known as _____taxonomy_____.

3. While many scientists have contributed to the study of taxonomy, one in particular can be credited for a tremendous amount of early work. His name was __Carl Linneas __.

4. Likely, the most important reason living things are placed into groups and given specific names, is that scientists need to _____communicate___ accurately about what they learn about living things with each other.

5. Living things all fall into one of five major groups known as kingdoms. These five kingdoms are: _____Animalia_____, ____Plantae____, _____Fungi____, _____Protista_____ and _____Moneran_____.

6. Examples of the Animalia kingdom might include: ____any animal_____, _____ and _____.

7. Examples of the Plantae kingdom might include: ___any plant_____, _____ and _____.

8. Examples of the Fungi kingdom might include: ____molds, mildew, mushroom, athlete's foot, ringworm_____, _____ and _____.

9. Bacteria and viruses are members of the _____Moneran_____ kingdom.

10. Unicellular, swimming creatures are members of the _____Protista_____ kingdom.

11. The classification division immediately beneath the kingdom level is the _phylum_____ level.

12. Humans fall into the phylum _____chordata_____ because of the presence of a _____spinal cord_____.

13. Why do dogs also belong to this phylum? _____presence of spinal cord

14. Beginning with kingdom and phylum, list the remaining levels of classifications for living things: kingdom, phylum, ____class_____, ___order_____, ___family____, ___genus_____ and _____species_____.

15. The level of classification which serves as the "first name" of an organism is _genus__ while the

___species__ level serves as the "last name."

16. There are rules for writing the scientific name of organisms. The ___genus_____ portion is capitalized while the __species____ portion is not. If hand-written, they are to be ____underlined_____ and if typed, ____italicized___.

17. In this lesson, we found that size was a determining factor used in grouping in members of the cat family. In horses and cattle, a determining feature was the number of _toes/hooves__ on each foot. Cattle are classified as being ___artiodactyles____ while horses are classified as being ___perissodactyles__.

18. How would a whitetail deer be classified? _____artiodactyles_____. How about a grey zebra? ____perissodactyles_____

19. Using the Catalogue of Life resource, find the scientific names for the following living organisms:

    A. Eastern Cottontail rabbit: _____*Sylvilagus floridanus* _____

    B. White oak: _____*Quercus alba* _____

    C. Oyster mushroom: _____*Pleurotus ostreatus* _____

    D. Domestic dog: _____*Canis familaris*_____

    E. House fly: _____*Musca domestica*_____

    F. Red squirrel: _____*Sciurus vulgaris* _____

20. Using the Catalogue of Life resource, find the common name for the following living organisms:

    A. *Bos taurus* _____Domestic cattle_____

    B. *Sus scrofa* _____Pig_____

    C. *Canis lupus* _____Wolf_____

    D. *Julglan nigra* _____Black Walnut_____

    E. *Homo sapien* _____Human_____

Name _____Date_____

Lesson 14   Practice Page 2

Instructions: Below, you will find clues to solve this crossword puzzle. Refer back to the text portion of the lesson for help.

Across

3. Kingdom that we find puffballs. Fungi

4. Fellow who did much of early taxonomy.

Carl Linneas

6. Kingdom of tiny swimmers. Protista

8. Chordata is an example Phylum

11. All members of the phylum Chordata have __backbones__

13. Lions and tigers and bears, oh my! Animalia

15. study of groupings of living things based upon similarities and differences   Taxonomy

Down

1. "highest" category of taxonomic hierarchy Kingdom

2. Sunflowers belong to this kingdom. Plantae

5. Species name is written using _lower case_____

7. Scientific name format when type-written. Italics

9. Genus name is always written using __upper case__

10. Kingdom that we find bacteria. Moneran

12. Phyla are divided into __classes___.

14. study of forms and shapes of living things Morphology

Name _____ Date_____

Lesson 15   Practice Page 1

Match the word or phrase in Column A with the <u>best</u> word or phrase in Column B. Write the letter from Column B beside the number in Column A. The first question has been completed for you!

Column A	Column B
__A__1. Kingdom Animalia	A Includes all animals
___H__2. Rotifera	B Phylum Platyhelminthes
___N__3. Phylum Mollusca	C Phylum Porifera
___J__4. Class Cestoda	D Phylum Cnidaria
___D__5. Characterized by having namatocysts	E Class Scyphozoa
___F__6. Sea anemones belong in this class	F Class Anthozoa
___G__7. Flukes are a member of this class.	G Class Trematoda
___B__8. Flat worms	H Have a little crown of twirling cilia.
___I__9. Comb jellyfish belong to this phylum.	I Phylum Ctenophora
___E__10. True jellyfish	J Tapeworms
___S__11. Phylum Annelida	K Spiny-headed worms
___L__12. Phylum Bryozoa	L Moss animals
___M__13. Phylum Brachiopoda	M Upper and lower shell
___C__14. Sponges	N Have a mantle which creates a shell
___P__15. Class Bivalvia	O Includes snails, slugs and whelks.
___O__16. Class Gastropoda	P Clams, oysters and scallops
___Y__17. Class Diplopoda	Q Members include the squid and octopus
__R___18. Chitons	R Shell made of eight valves known as plates
___K__19. Phylum Acanthocephala	S Segmented worms like earthworms
___X__20. Class Insecta	T Segmented body parts and jointed legs.
__U___21. Class Arachnida.	U Spiders
___V__22. Subphylum Crustacea	V Two pairs of antennae and ten legs.
___W__23. Class Chilopoda	W Centipedes
___Q__24. Class Cephalopoda	X Insects
___T__25. Phylum Arthropoda .	Y Millipedes
___Z__26. Phylum Echinodermata.	Z Pentaradially symmetrical.
__AA___27. Phylum Hemichordata.	AA Acorn worms
__BB___28. Phylum Chordata	BB Animals with backbones.

Name _____ Date_____

Lesson 16   Practice Page 1

Match the word or phrase in Column A with the <u>best</u> word or phrase in Column B. Write the letter from Column B beside the number in Column A. The first question has been completed for you!

Column A	Column B
__A__ 1. Phylum chordate	A Phylum of animals possessing a notochord, gill slits and a postanal tail (at one point in development).
__B__ 2. Subphylum Cephalochordata	B Lancelets
__D__ 3. Class Chondrichthyes	C No-jawed fish; lamprey and hagfish
__J__ 4. Order Trachystoma	D Sharks, rays and skates
__E__ 5. Class Osteichthyes	E Fishes that have bony skeletons
__I__ 6. Order Anura	F Aquatic at some point in their lives and then become terrestrial
__F__ 7. Class Amphibia	G Amphibians with no legs, blind worms
__G__ 8. Order Apoda	H Newts and salamanders
__C__ 9. Class Agnatha	I Amphibians with no tail, frogs, toads.
__P__ 10. Class Aves	J Rough-mouthed amphibians
__K__ 11. Class Reptilia	K Covered with scales, only have lungs, internal fertilization
__L__ 12. Order Rhynchocephalia	L Tuatara of New Zealand
__V__ 13. Order Edentata	M Order include turtles and tortoises.
__O__ 14. Order Squamata	N Includes crocodiles and alligators
__U__ 15. Order Chiroptera	O Snakes and lizards
__N__ 16. Order Crocodilia	P Class includes all birds
__H__ 17. Order Urodela	Q Warm-blooded, milk-producing, hairy animals
__T__ 18. Order Insectivora	R Included the egg-laying mammals: platypus and echidna
__Q__ 19. Class Mammalia	S Pouch-possessing mammals
__M__ 20. Order Chelonia	T Moles and shrews
__FF__ 21. Order Rodentia	U flying mammals, bats
__R__ 22. Order Monotremata	

___W__25. Order Cetacea

___Z__26. Order Carnivora

___X__27. Order Sirenia

___CC__28. Order Artiodactyla

___Y__29. Order Proboscidae

__EE___30. Order Lagomorpha

__AA___31. Order Pinnipedia

__BB___32. Order Perissodactyla

__DD___33. Order Primates

V.   Peg-like teeth armadillo and sloth

Includes whales, porpoises and dolphins.

Members of this order include the manatees

Elephants

Meat-eaters; foxes, wolves, dogs, cats

Seals, walruses

Horses, donkeys, rhinos

Include the cattle, sheep, deer, giraffe

Monkeys, lemurs, gibbons, orangutans, chimpanzee, gorillas and humans.

Includes rabbits, hares and the pikas

Rats, mice, squirrels

The pangolins

Name _____Date_____

Please fill in the blank(s) for each statement below. Use your lesson to help you.

1. Plants, through the process of ___photosynthesis__ are able to take carbon dioxide, water and sunlight to produce _____glucose_____ which is fuel for the plant.

2. Plants are considered to be _____autotrophs____ in that they are able to produce their own food supply. Living things which depend upon other living things as food are said to be ____heterotrophs____.

3. Photosynthesis in plants takes place in the ___chloroplasts___ of the cells of plants.

4. Unlike the cells of animals, plants have ___cell walls__ which are made up primarily of __cellulose_ which are long chains of glucose molecules.

5. Taxonomists divide plants into two large groups based upon the presence of __vascular__ tissue.

6. Vascular tissue which carries water is known as the ___xylem___ while vascular tissue which carries food (glucose) around the plant is known as ____phloem_____.

7. Instead of being divided into phyla, plants within the Kingdom Plantae are divided into ___divisions___.

8. There is one division of plants in the non-vascular division. This is the Division ___Bryophyta___.

9. Three classes are present within this division. The plants found in these classes include the _____mosses_____, _____hornworts___ and _____liverworts____.

10. The Division Psilophyta has ____fern_____-like leaves. The most common member is the _whisk fern_.

11. The Division Sphenophyta includes plants known as _____horsetails____ which have hollow stems and contain ___silica_____.

12. The Division Pterophyta includes __ferns____ which have well developed underground rhizomes.

13. Cycads belong to the Division ____Cycadophyta_____ and have seeds not enclosed in ___fruit___.

14. The lone member of the Division Gingkophyta is the ____gingko tree_____.

15. The _____welwitschia____ plant is a member of the Division Gneophyta has two very long leaves and a very short trunk.

16. Trees such as pines, firs and spruces belong to the Division ____Coniferophyta___. They are com-

monly referred to as ___evergreens_____ and have their seeds in _____cones_____.

17. Their leaves are like __needles___.

18. Plants which have flowers and seeds enclosed in a fruit belong to the Division ____Anthophyta_____.

19. There are two classes found within the Division Anthophyta. The first includes plants with one cotyledon and are commonly referred to as _____monocots_____. Examples of these include: ___corn, grasses, irises, tulips_____

20. Flowering plants with seeds having two cotyledons are commonly referred to as ____dicots_____ and examples are: ____beans, peas, legumes_____.

21. Flowers with both male and female structures present are known as ___perfect__ flowers whereas flowers with only one sex of reproductive structure are known as ____imperfect___ flowers.

22. The male parts of a flower are the stem-like ____stamens____ with the pollen-producing part known as the ____anther_____.

23. The female part of a flower includes the ____pistil____ which has a stem-like portion known as the ____style_____ and the sticky top known as the _____stigma_____.

24. Pollen carry two types of cells: one which forms the _____pollen tube___ while the second divides to create two ___sperm cells_____.

25. The pollen tube allows the ____sperm_ cells to join the ____ovules__ present in the ovary of the flower. From this joining process, many cell divisions result and the seed, enclosing the plant embryo, develops.

Name _____ Date_____

Lesson 17   Practice Page 2

Use the clues on the next page to solve this crossword puzzle regarding the plant kingdom.

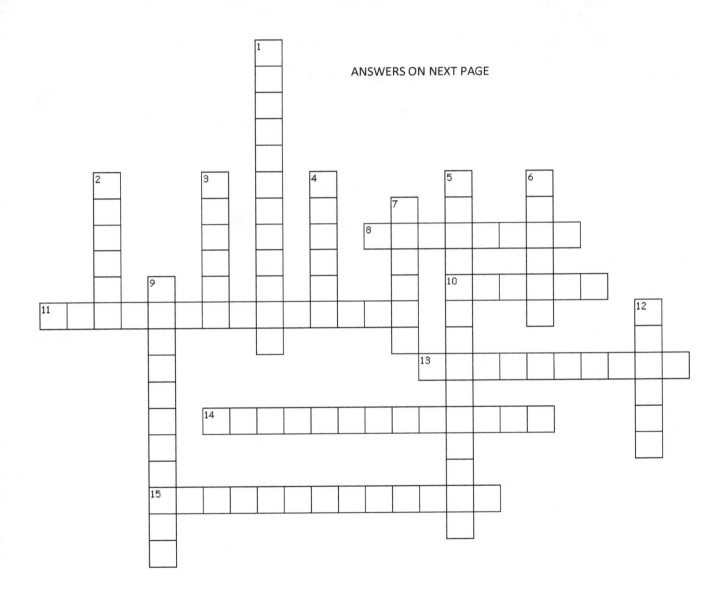

ANSWERS ON NEXT PAGE

Clues:

Across

8. plants having parallel leaf veins and seeds with one cotyledon MONOCOTS

10. structures within ovary which are female contribution to new plant embryos   OVULES

11. tissue of plants which is capable of transporting water and food products VASCULAR TISSUE

13. division of plants which includes flowering plants ANTHOPHYTA

14. division of plants which includes pines, spruces and firs CONIFEROPHYTA

15. living things which depend upon other living things as a food supply HETEROTROPHIC

Down

1. organelles which carry out photosynthesis CHLOROPLASTS

2. plants with branched leaf veins and seeds with two cotyledons DICOTS

3. sticky end of female pistil where pollen lands STIGMA

4. leaf-like covering of a flower SEPALS

5. process whereby plants create glucose from sunlight, water and carbon dioxide PHOTOSYNTHESIS

6. colorful part of flower which attracts pollinators PETALS

7. particle which contains cells which create pollen tube and sperm cells POLLEN

9. plants are considered to be this because they can make their own food AUTOTROPHIC

12. portion of male flower part that produces pollen grains ANTHER

Name _____Date_____

Lesson 18   Practice Page 1

Please fill in the blank(s) for each statement below. Use your lesson to help you.

1. Members of the Kingdom Monera are considered to be _____prokaryotes____ meaning they have no nuclear membrane nor organized organelles.

2. Members of the Kingdom Monera also are _____unicellular____ meaning they solely consist of being one cell.

3. The primary forms of reproduction in the Kingdom Monera is through _____binary fission___ and _____budding_____.

4. The phylum of Monera which includes the bacteria which causes disease is the Phylum _____Schizophyta_____.

5. Members of this phylum (question 4) live off of _____dead living things____ or live as parasites of living things.

6. Viruses consist of a strand of ____DNA__ or ____RNA___ enclosed by a _____protein coat_____ known as the _____capsid_____.

7. Viruses are considered to be ____obligate___ __intracellular organisms_____ as they must live within the cells of a host to survive.

8. Control of viruses is mainly through the use of __vaccines___ which stimulate the body's immune system to produce ___antibodies/immunoglobulins___ to fight off the disease when encountered at a later time.

9. Viruses work by entering a host cells and taking over ___control_ of the cell's DNA in an effort to get the cell to make more viruses. The host cell then dies.

10. Viruses are classified as either being DNA or ____RNA___ viruses.

Name _____Date_____

Use the clues below to solve this crossword puzzle regarding the Kingdom Moneran.

Clues:

Across

3. "living" things which cause disease VIRUSES

4. agent given to stimulate immune system to fight off virus should it ever arrive   VACCINE

6. have no cell membrane nor organized organelles   PROKARYOTES

7. meaning only one-cell-big UNICELLULAR

Down

1. means of asexual reproduction utilized by monerans   FISSION

2. kingdom of unicellular, prokaryotes   MONERAN

5. protein covering of a virus   CAPSID

Name _____Date_____

Lesson 19   Practice Page 1

Please fill in the blank(s) for each statement below. Use your lesson to help you.

1. Members of the Kingdom Protista are all _____unicellular_____ creatures meaning they are made up of organisms of only one cell.

2. They differ from members of the Kingdom Moneran in that they have _____membrane bound_____ organelles and are therefore identified as being _____eukaryotic_____ as opposed to the Monerans which are prokaryotic.

3. While there are several phyla of the Kingdom Protista, these can be grouped into three main groups. These are the ___animal__-like protists, the ____plant___-like protistas and the ___fungal__ -like protists.

4. Members of the animal-like protists are grouped based upon ___locomotion___.

5. ___Pseudopodia_____ are the false-foot like appendages of Phylum Sarcodina.

6. _____Cilia_____ are the hair-like projections found in the Phylum __Cililphora_.

7. Members of the Phylum _____Zoomastigina_____ move about through the use of a flagellum which is a tail-like structure capable of propelling the organism through the water.

8. Members of the Phylum _____Sporozoa____ are animal-like yet don't move about. Members in-clude the _____malarial organisms and Toxoplasma organisms_____.

9. The fungal-like protists include slime molds which move about on the ___forest floor___ eating up dead leaves.

10. The devastating potato famine of the 1800's was caused by a member of the ___Oomycota___ phylum.

11. The plant-like protists include various phyla of _____algae____. They are categorized based up-on the presence of various colors of _____pigments_____ as well as how they store _____glucose_____.

12. Chlorophyta are the ___green__ algae. Rhodophyta are the _____red___ algae.

13. Phaophyta are the _____brown____ algae. Chrysophyta are the _____golden brown___ algae.

14. Pyrrophyta have the ability to produce light through a process known as __bioluminescence__.

Lesson 19   Practice Page 2

Use the clues on the next page to solve this crossword puzzle regarding the Kingdom Protista.

ANSWERS ON NEXT PAGE

Clues:

Across

2. algae capable of producing light PYRROPHYTA

5. animal-like protists PROTOZOANS

7. fungal-like protists of forest floor   SLIME MOLDS

10. green algae CHLOROPHYTA

11. brown algae PHAEOPHYTA

14. hair-like projection allowing for movement CILIA

Down

1. living thing capable of producing light BIOLUMINESCENT

3. red algae RHODOPHYTA

4. false foot PSEUDOPDIA

6. golden brown algae   CHRYSOPHYTA

8. having membrane-bound organelles EUKARYOTES

9. tail-like structure of protozoans FLAGELLUM

12. plant-like protists ALGAE

13. *Phytophthora spp.* caused this in Ireland in the mid-1800s
POTATO FAMINE

Name _____Date_____

Lesson 20   Practice Page 1

Please fill in the blank(s) for each statement below. Use your lesson to help you.

1. Fungi are ____saprophytes___ meaning they live off of dead organisms. They are known as ____decomposers____.

2. Can be categorized according to the presence or lack of _____crosswalls___ in their hyphae. Hyphae can be considered to be like __roots___ which grow through the substance being "eaten" by the fungus.

3. Fungi can reproduce __sexually__ and _____asexually_____.

4. Fungi which belong to the Division Zygomycota have no ____crosswalls__ in their hyphae. The common _____black bread mold_____ is a member of this division.

5. Fungi which belong to the Division Basidiomycota have ____crosswalls__ in their hyphae. Familiar members include _____mushrooms_____.

6. Fungi which belong to the Division Ascomycota have crosswalls with _____holes or perforations_____ and reproduce through the use of __asci___or sacs which produce __ascospores____. Morels and ___bread yeasts__ are members of this division.

7. Fungi which belong to the Division Deuteromycota have _____reproductive_____ cycles which are not yet fully understood. Athlete's foot fungus and the fungus used to create _penicillin____ and flavor ___cheeses_____ belong in this division.

Name _____Date_____

Lesson 20   Practice Page 2

Use the clues below to solve this crossword puzzle regarding the Kingdom Fungi

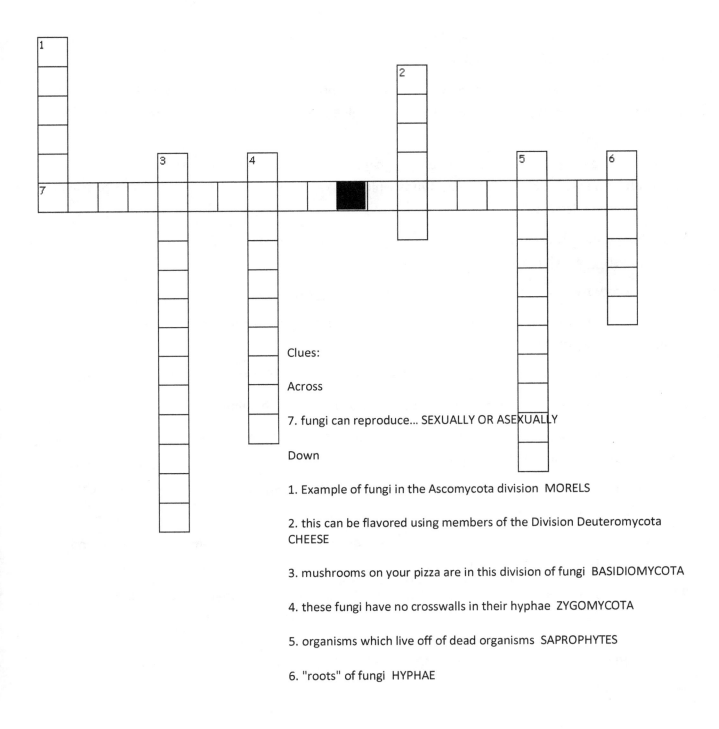

Clues:

Across

7. fungi can reproduce... SEXUALLY OR ASEXUALLY

Down

1. Example of fungi in the Ascomycota division MORELS

2. this can be flavored using members of the Division Deuteromycota CHEESE

3. mushrooms on your pizza are in this division of fungi BASIDIOMYCOTA

4. these fungi have no crosswalls in their hyphae ZYGOMYCOTA

5. organisms which live off of dead organisms SAPROPHYTES

6. "roots" of fungi HYPHAE

Name _____Date_____

Lesson 21   Practice Page 1

Please fill in the blank(s) for each statement below. Use your lesson to help you.

1. The study of how living things are made is known as _____anatomy__ while how living things func-tion, is known as ____physiology_____.

2. At birth, a baby has over ___300_ bones of which several fuse together to reduce the number down to about ____206____ in an adult.

3. There are the __long__ bones which consist of the bones making up the arms and legs, the _flat_ bones which are bones primarily of the skull and pelvis and the __irregularly-shaped__ bones which are the bones of the spinal column

4. Initially, the long bones were formed of ___cartilage__ which over time is replaced by __calcium__ and __phosphorus__ containing compounds which make the bone hard. This hardening process is called __ossification__.

5. On long bones, there are two __epiphyses___, one found on each end and then one ___diaphysis__ which joins the two epiphyses.

6. Locations where ossification begins are known as ____centers of ossification_____.

7. The boundary where growth of the bone occurs is referred to as the __growth plate__ or _epiphyseal line_ of the bone.

8. The joints between the long bones of the arms and legs are known as ___synovial___ joints.

9. If you look at the ends of long bones, you'll find a layer of ___hyaline_____ cartilage.

10. To reduce friction between the ends of two adjoining bones, a fluid known as ___synovial fluid__ keeps the bones gently sliding over each other

11. The tough membranes which bridge the joint space between the two bones is known as the __synovial capsule__

12. Over time and with age, this slippery joint surface may become worn and result in pain and in-flammation. This is known as __arthritis___.

13. The joint between the femur and the pelvis at the hip is classified as a __ball and socket___ joint.

14. The joint at the knee and elbow are classified as ____hinge___ joints as the mainly allow for bending in one direction.

15. The cells which make up bone are known as __osteocytes___.

16. There are two kinds of osteocytes. The ___osteoblasts___ are bone cells which have the responsibility of building new bone, like along the growth plate, while the ____osteoclasts__ have the job of taking apart or tearing-down bone.

17. Bones are also very important when it comes to ___protection__ of vital organs in the body. The flat bones of the skull protect the ____brain___ while the irregularly-shaped bones of the spinal column protect the __spinal cord__. The ____ribs_____ also are very important for protecting the lungs, heart and upper abdominal organs.

18. The production of red blood cells takes place in the __marrow cavity__ of bone This tissue is known a hemopoietic tissue which literally means blood-making tissue. The marrow cavity is also a location for ____fat storage_____ in adults.

19. Bones also function as a storage site or depot for the element ___calcium____. Not only is calcium important for the ____hardness____ of our bones, calcium plays a major role in our __nervous system___.

20. There are three types of muscle in the body. There is ___smooth muscle___ which is found in our internal organs like our stomach, intestines, bladder and respiratory tract. Then there is __cardiac muscle_ which is specifically made for our heart. Finally, there is ____skeletal/striated____ which enables us to move.

21. Skeletal muscle is also referred to as ____striated_____ muscle.

22. There are components of skeletal muscle cells made up of what are known as ___contractile__ proteins or filaments. There are two types of contractile proteins: _____actin_____ and ____myosin_____.

23. The attachment of muscles to bones is made through tough connective tissue material known as ____tendons____.

24. Connective tissue structures which connect bones to bones are known as __ligaments____.

25. Muscles can only actively ___contract____.

26. The end of the muscle which is nearer to the main trunk (chest and abdomen) of the body is known as the ___origin____ of the muscle. The opposite end of the muscle which is farther from the main trunk of the body is known as the _____insertion____ of the muscle.

27. Bending of a joint is referred to as ____flexion____ while straightening a joint is known as ___extension of___ the joint.

28. Inward movement is referred to as ____adduction____ and outward motion is referred to as ___abduction__.

29. Cells of the nervous systems are called ____neurons__ and consist of a central nucleus-containing area known as the _____soma_____. A single _____axon_____ is present with multiple ___dendrites___.

30. An action potential (stimulus) travels into the _____dendrite____ "end" and out through the _____axon____ "end" of a neuron.

31. Axons terminate on ____myocytes__ (muscle cells) where ____neurotransmitters__ are released to activate muscle contraction. The most common neurotransmitter is __acetylcholine__ (ACH).

32. Neurons which carry messages out to muscles are known as ____efferent____ or _____motor___ neurons.

33. Neurons which carry messages inward from sensory organs such as the skin or eyes are called ___afferent____ or _____sensory_____ neurons.

34. Conduction of a stimulus is the result of _____sodium____ and _____potassium_____ being pumped in and out of the neuron. The element _____calcium___ mediates this process.

35. ___Anesthesia_ interferes with transmission of nerve impulses or neurotransmitter function which results in the desired control of pain and/or movement.

Name _____Date_____

Lesson 21   Practice Page 2

Use the clues on the next page to solve this crossword puzzle regarding the skeletal, muscular and nervous systems.

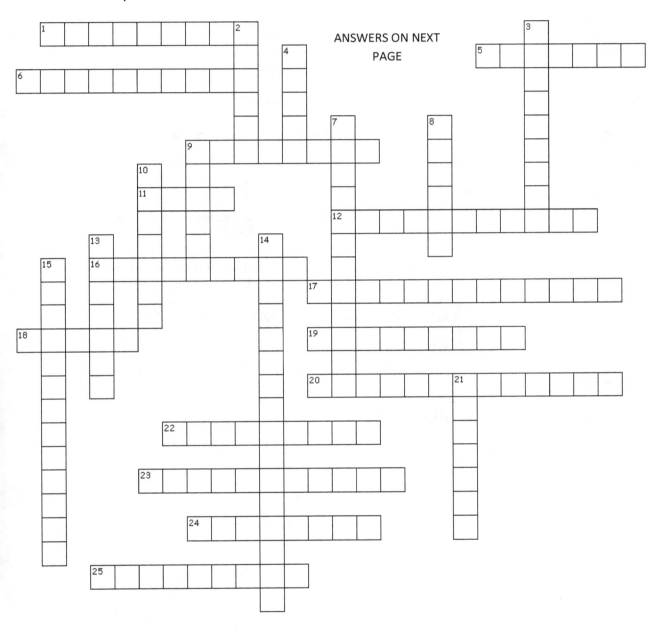

ANSWERS ON NEXT PAGE

Clues:

Across

1. bones of arms and legs LONG BONES

5. making joint angle less FLEXION

6. type of osteocyte which breaks down bone OSTEO-CLAST

9. type of muscle which moves bones SKELETAL

11. "end" of neuron where stimulus departs AXON

12. type of osteocytes which builds bone OSTEOBLASTS

16. ends of long bones EPIPHYSES

17. hip joint BALL AND SOCKET

18. type of neuron which results in muscle action MOTOR

19. part of bone between epiphyses DIAPHYSIS

20. type of joint between long bones SYNOVIAL

22. making joint angle greater EXTENSION

23. contractile proteins of muscles ACTIN MYOSIN

24. only intentional action of a muscle CONTRACT

25. bones of skull and pelvis FLAT

21. study of parts of body ANATOMY

Down

2. type of joint between flat bones SUTURE

3. "end" of neuron where stimulus arrives, multiple DENDRITES

4. elbow joint HINGE

7. location for red blood cell production MARROW CAVITY

8. nerve cell NEURON

9. type of muscle of internal organs other than heart SMOOTH

10. element stored in bones CALCIUM

13. type of neuron which carries sensory information back to brain for interpretation   SENSORY

14. chemicals which move from end of nerve to muscle to trigger contraction NEUROTRANSMITTER

15. lubricant of joints SYNOVIAL FLUID

Name _____Date_____

Lesson 22   Practice Page 1

Please fill in the blank(s) for each statement below. Use your lesson to help you.

1. The primary function of the circulatory system is to deliver _____oxygen_____ and ___glucose___ to all cells of the body.

2. _____Capillaries_____ are the smallest blood vessels, are only __one____ cell layer in thickness and have windows or openings known as ____fenestrations___ to allow certain substances to pass.

3. The __size____ of the fenestrations of capillaries varies throughout the body with some areas be-ing very tightly controlled and other areas like in the _____spleen____ where whole blood cells freely move in and out of the capillaries.

4. Capillaries are "fed" by ____arterioles_____ which in turn receive blood from _____arteries_____ which get blood from the largest artery of all which is the _aorta__.

5. In the human heart there are four _____chambers_____ . The two which receive blood from out-side the heart are known as the _____atria_____ while the two which pump blood out of the heart are known as the _____ventricles_____.

6. Blood from the body that enters the heart is ___low___ in oxygen content and _high__ in carbon dioxide. The chamber of the heart it enters is the _____right____ atrium.

7. From this chamber, upon contraction of the right atrium, blood moves down through the _____right AV_____ valve into the _____right_____ ventricle.

8. Upon contraction of the right ventricle, blood moves out of the heart destined for the __lungs_ to get ____oxygen____ and leave behind _____carbon dioxide____.

9. Blood returning from the lungs is ____high___ in oxygen and ____low___ in carbon dioxide. It en-ters the heart in the _____left____ atrium.

10. From the left atrium, blood moves downward through the _____left AV____ valve into the ____left____ ventricle.

11. Upon contraction, blood leaves the left ventricle passing through the _____aortic___ valve heading to the _____body____ through the largest artery known as the ____aorta_____.

12. The right and left atrioventricular valves consist of flap-like structures known as the __cusps__ of the valves. Preventing these structures from prolapsing upward into the atria above are tough string-like structures known as _____chordae tendonae_____.

13. If a valve does not close properly which allows blood to "leak" in an unintended direction or is too "tight" (which is known as ____stenosis___,) one can hear abnormal sounds known as ____murmurs____.

14. Air moves down from the nose and mouth through the _____trachea_____ which branches into _____bronchi_____ which branch further into _____bronchioles_____.

15. The bronchioles end in small grape-like clusters of air sacs known as _____alveoli_____.

_____Capillaries_____ wrap around the alveoli to pick up oxygen and leave behind _____carbon dioxide_____.

16. In the body, capillaries returning from cells deliver blood into the smallest veins known as _____venules_____.

17. Unlike arteries, veins have ____thin_____ walls and have small _____valves____ within them to prevent backflow of blood.

18. The watery portion of blood is known as ___plasma/serum____.

19. Red blood cells are _____erythrocytes_____ while the scientific name for the group of white blood cells are the _____leukocytes_____.

20. Red blood cells are produced in the _____bone marrow_____ and have the primary job of _____carrying oxygen_____ through the work of a special protein found within them known as _____hemoglobin_____.

21. The decrease in the functioning number of red blood cells is known as ____anemia__.

22. ____Leukocytes__ function as part of our body's defense system against invading organisms.

23. _____Lymphocytes____are white blood cells which create antibodies or __immunoglobulins__ which work as ___bullets/spears/knives etc.___ for later attack of invaders.

24. __Eosinophils_, __basophils__ and ___mast cells_ work to mediate inflammation in the body.

25. Carbon monoxide causes problems in our body because it replaces __oxygen__ being carried by ____hemoglobin_____ within red blood cells.

Name _____ Date_____

Lesson 22   Practice Page 2

Use the clues below to solve this puzzle.

Across

3. tiny veins VENULES

6. smaller than an artery but larger than a capillary ARTE-RIOLES

7. flap portion of valve CUSP

8. most abundant leukocyte NEUTROPHILS

9. chambers of heart which pump blood out of heart VENTRICLES

10. location of red blood cell production   MARROW

11. necessary substance for respiration   OXYGEN

13. oxygen-carrying protein of red blood cells HEMO-GLOBIN

17. low number of erythrocytes   ANEMIA

20. red blood cells   ERYTHROCYTES

21. watery portion of blood PLASMA

Down

1. smallest blood vessel CAPILLARY

2. "strings" of valve which prevents valvular prolapse CHORDAE TENDONAE

4. white blood cells which makes antibodies LYMPHO-CYTES

5. windows of capillaries FENESTRATIONS

12. "gate" within heart which allows for one direction flow VALVE

14. white blood cells LEUKOCYTES

15. fuel of body carried by blood GLUCOSE

16. largest artery of all AORTA

18. chambers of heart which welcome blood ATRIA

19. pumper of blood HEART

Name _____Date_____

Lesson 23   Practice Page 1

Please fill in the blank(s) for each statement below. Use your lesson pages to help you.

1. The main function of the digestive system is to take ____food_____ and convert it into glucose to be delivered to _body____ by the ___circulatory_____ system.

2. A bite of food is known as a _____bolus_____.

3. Saliva, which contains _____enzymes/amylase_____, is mixed with the bolus to begin the digestion process.

4. Food moves from the mouth to the stomach through the _____esophagus____ by the action of _____smooth_____ muscles.

5. The stomach mixes the food with substances produced by the chief and parietal cells. The chief cells create _____pepsinogen_____ which reacts with hydrochloric acid from the ____parietal cells___ to produce pepsin, a ____proteolytic___ enzyme.

6. The "front door" of the stomach is the ___cardiac sphincter__ while the "back door" of the stomach known as ___pyloric sphincter_____.

7. The small intestine is divided into three segments: the ____duodenum___ which is closely associated with the pancreas, the ____jejunum____ where microvilli allow for great surface area for nutrient absorption and the ____ileum_____.

8. The pancreas secretes the hormone ___insulin__ which regulates the ability of glucose to enter ____glucose___ of the body.

9. The ____liver_____ is known to have close to 500 specific jobs in the body which includes the __filtering of blood_____ arriving from the small intestine as well as the manufacture of ____enzymes_____ and _____glucagon____ (the storage form of glucose.)

10. At the junction of the ileum and the large intestine (colon) is the __cecum___ or ___appendix___.

11. The primary role of the large intestine is ___water reabsorption_____ and the production and absorption of __vitamin K___ which is vital for blood __clotting___.

Name _____Date_____

Lesson 23   Practice Page 2

Use the clues below to solve this crossword puzzle about the digestive system.

Across

2. this sphincter is the back door PYLORIC

6. process of taking food and converting to glucose DIGESTION

8. bite of food BOLUS

9. pancreas secretes this hormone which allows glucose to enter cells INSULIN

10. these are found in saliva to begin process of digestion ENZYMES

11. colon reabsorbs this WATER

13. first section of small intestine with pancreas as a pal DUODENUM

14. allow for great surface area within small intestine MICROVILLI

16. fancy name for spit SALIVA

Down

1. enzymes which lyse proteins PROTEOLYTIC ENZYMES

2. cells of stomach which produce hydrochloric acid PARIETAL CELLS

3. cells of stomach which produce pepsinogen CHIEF CELLS

4. tube from mouth to stomach, made of smooth muscle ESOPHAGUS

5. longest section of small intestine JEJUNUM

7. large intestine produces and absorbs this vitamin VITAMIN K

12. blind tube portion of intestine APPENDIX

15. this sphincter is the front door CARDIAC

Name _____Date_____

Please fill in the blank(s) for each statement below. Use your lesson pages to help you.

1. The ___renal_____ and _____urinary_____ systems of our body work to remove wastes created by cells as well as cell remnants from the body.

2. The renal system consists of our two ____kidneys____ and blood vessels which carry "dirty" blood to and "clean" blood away from the _____kidneys_____.

3. The filtration unit of the kidney is the ___glomerulus___.

4. Waste products as well as water can leave through ____fenestrations___ in the walls of the arterioles within the glomerulus, however, ___water_____, ____sodium_____ and __potassium____ are returned back to circulation and therefore conserved.

5. The kidneys also maintain _____pH_____ of the blood.

6. The kidneys produce hormones which control ___blood pressure___ and the creation of ___erythrocytes/red blood cells___.

7. The presence of ____proteins____ in the urine indicates possible kidney damage or damage to other parts of the urinary system such as the ____bladder_____ or ____urethra____.

8. Glucose in the ___urine_____ indicates it has exceeded the threshold capacity of the kidney.

___Insulin___, produced by the ___pancreas____, functions to open "doors" of cell membranes to allow glucose to enter.

9. Lack of insulin can result in ____high___ levels of blood glucose, presence of glucose in the ____urine____ and "starving" _____cells_____ all over the body.

10. Urine leaves the kidneys through the _____ureters____ which drain into the _urinary bladder__.

11. When full, the bladder is emptied to the exterior of the body through the ___urethra__.

Name _____Date_____

Lesson 24   Practice Page 2

Use the clues below to solve this crossword puzzle about the renal and urinary systems.

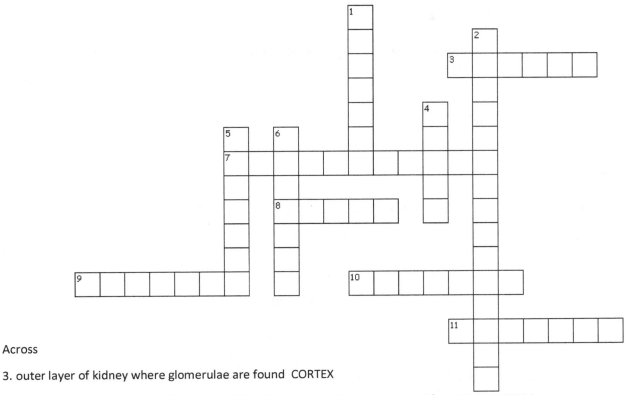

Across

3. outer layer of kidney where glomerulae are found CORTEX

7. body system which includes kidneys and blood vessels carrying blood to and fro RENAL SYSTEM

8. waste-full liquid produced by kidneys URINE

9. inner layer of kidney MEDULLA

10. presence of this in urine means too much is in blood GLUCOSE

11. we have two and their job is to filter blood, regulate pH and blood pressure and RBC production KIDNEYS

Down

1. these tubes carry urine from kidney to urinary bladder URETERS

2. these two elements are also "saved" by the kidneys and not flushed away SODIUM POTASSIUM

4. this gets conserved by kidneys so urine is not too watery   WATER

5. this tube empties bladder to exterior of body URETHRA

6. this hormone produced by pancreas opens "doors" of cells to welcome glucose INSULIN

Name _____Date_____

Please fill in the blank(s) for each statement below. Use your lesson pages to help you.

1. The _____endocrine_____ system is the body system which regulates the activities of many other body systems through the function of _____glands____.

2. There are __2__ types of glands: __exocrine___ which have ducts or tubes in which substances are exported and ____endocrine_____ glands which deposit substances directly into the circulatory system.

3. Levels of hormones in the body are maintained using a _negative feedback mechanism__ whereby low levels trigger the endocrine gland to begin production of the hormone.

4. The ____pituitary gland___ is known as the "master" gland as it produces hormones which affect many organs in the body as well as other _____endocrine___ glands which in turn affect body organs.

5. The ____anterior___ pituitary gland produces thyroid stimulating hormone (TSH), adrenocortico-tropic hormone (ACTH), human growth hormone (HGH), prolactin and the gonadotropic hormones.

6. TSH affects rate of ___metabolism__ which is how fast or slow cells utilize __glucose__.

7. ACTH signals the ___adrenal cortex__ to produce anti-inflammatory substances and water conser-vation hormones.

8. HGH regulates __growth____ in the body.

9. Prolactin regulates development of _____mammary/breast____ tissues as well as milk production in women.

10. The gonadotropic hormones in women stimulate the _____ovaries_____ to produce _____ova_____ and then maintain pregnancy.

11. The gonadotropic hormones in men stimulate the _____testis_____ to produce ____sperm cells_____ and ____testosterone___.

12. Hormones of the posterior lobe of the pituitary gland include ____vasopressin_____ also known as _____ADH_____ which regulates water conservation by the kidneys and oxytocin which stimu-

lates the _____uterus_____ to contract during birth and milk _____let-down____ while nursing.

13. The parathyroid glands control _____calcium____ levels in the body.

14. The ___pineal_____ gland regulates daily sleep and wakefulness patterns.

15. The thymus, while only active in children, functions to produce _____killer T lymphocytes___ which function throughout life to defend against ___disease/invaders___.

Name _____Date_____

Lesson 25   Practice Page 2

Use the clues on the next page to solve this crossword puzzle about endocrine system.

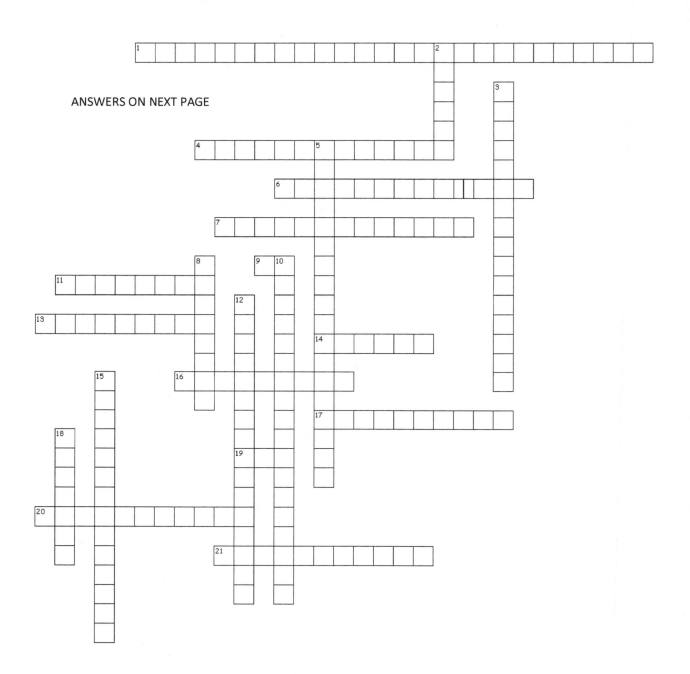

ANSWERS ON NEXT PAGE

Across

1. this hormone stimulates development of follicles on the ovary in women FOLLICE STIMULATING HORMONE

4. target organs of ACTH   ADRENAL GLANDS

6. the "master" gland   PITUITARY GLAND

7. target organs of TSH THYROID GLANDS

9. this hormone stimulates production of testosterone in men   LH

11. this hormone stimulates the uterus to contract during labor as well as milk let-down during nursing OXYTOCIN

13. this body system utilizes hormones to regulate other body systems   ENDOCRINE

14. target organs of gonadotropins in general GONADS

16. "back" side of pituitary gland POSTERIOR

17. rate of glucose use by cells METABOLISM

19. this hormone stimulates production of sperm cells in men FSH

20. this tiny gland regulates sleep and wakefulness patterns PINEAL GLAND

21. produced by the posterior pituitary gland to regulate water levels in the body VASOPRESSIN

Down

2. this childhood gland produces killer-T cells for lifelong defense strategies THYMUS

3. adrenal cortex produces these agents to reduce pain, swelling and redness ANTI INFLAMMATORY

5. this hormone stimulates ovulation in women LUTEINIZING HORMONE

8. "front" side of pituitary gland   ANTERIOR

10. HGH which regulates growth is short for _____ HUMAN GROWTH HORMONE

12. this mechanism is based upon low levels triggering action upon producing glands NEGATIVE FEEDBACK

15. these glands have tubes or ducts to delivery products to body EXOCRINE GLANDS

18. levels of this important element are controlled by the parathyroid glands CALCIUM

Name _____Date_____

Lesson 26   Practice Page 1

Please fill in the blank(s) for each statement below. Use your lesson pages to help you.

1. Sperm and ova are ___1N__haploid_____ cells having under gone meiosis. Sperm are formed in the _____testis_____ and ova form in the _____ovary_____.

2. The testis are located outside the body in the _____scrotum_____. The ovaries are located in the ____upper abdomen_____.

3. Sperm leave the testis through the _____epididymis_____ and continue through the vas deferens to the _____urethra_____.

4. Accessory sex glands which include the seminal vesicles, Cowper's gland and prostate gland provide __nutrients_____ and __pH adjusting___ substances to the sperm to create _semen_.

5. The ____penis_____ allows the man to deposit semen into the woman's _vagina___.

6. The cervix functions like a __door___ to open at specific times in response to _estrogen/hormones_ to allow sperm to enter or be closed during pregnancy.

7. The release of the ovum is known as ____ovulation__. The ovum is gathered by the _____infundibulum_____ and moved down the fallopian tube where ___fertilization_____ usually takes place.

8. The embryo moves down to the ____uterus_____ to implant and continue development.

9. Blood never mixes directly between the _____mother_____ and developing _baby__. Nutrients and wastes along with oxygen and carbon dioxide readily move across ____cell_____ membranes of the _____placenta_____.

10. Pregnancy in humans is _____nine_____ months.

11. ____Oxytocin_____ from the posterior pituitary gland stimulates the uterus to begin contractions while decreasing ____progesterone____ levels allows the cervix to dilate.

12. Should the baby be too large to deliver, a __Cesarean section____ may be performed to surgically remove the baby from the uterus.

Lesson 26   Practice Page 2

Use the clues below to solve this crossword puzzle.

Across

5. process of going from 2N to 1N MEIOSIS

7. length of time of pregnancy NINE MONTHS

11. tube leading from vas deferens to exterior of body URETHRA

12. location for ovum development OVARY

13. female gametes OVA

14. "door" of female reproductive system CERVIX

15. liquid carrying sperm cells SEMEN

16. location where embryo implants and grows throughout pregnancy UTERUS

17. condition of having one-half the total number of chromosomes 1N HAPLOID

18. usual location of fertilization FALLOPIAN TUBE

19. process where gametes join to create new 2N individual FERTILIZATION

Down

1. location of testis outside body SCROTUM

2. location for sperm and testosterone production TESTIS

3. includes seminal vesicles, Cowper's gland and prostate gland ACCESSORY SEX GLANDS

4. tube leading from testis to vas deferens EPIDIDYMIS

6. process where follicle ruptures to release ovum OVULATION

8. hormone which causes uterine contractions OXYTOCIN

9. male gametes SPERM

10. location where blood vessels from baby come close to blood vessels of mother but never actually join PLACENTA

Name _____Date_____

Lesson 27   Practice Page 1

Please fill in the blank(s) for each statement below. Use your lesson pages to help you.

1.   ____Seeing_____,   ____Hearing_____,   _____Smelling_____, ____Tasting_____, and _____Feeling/Touch_____ are our five means of sensing what it happening in our environment.

2.   The clear outer covering of our eyes is called the __cornea___ which transitions to the ___sclera___ which is the white portion on the perimeter of the eye.

3. Tears from the __lacrimal gland_ continually bathe the cornea to maintain hydration and wash away __debris___.

4. The colored portion within the eye is the ___iris____ and creates the circular _pupil___ which adjusts the amount of ___light____ which can enter the eye.

5. The fluid in front of the lens of the eye is the __aqueous___ humor while the fluid behind the lens, which is much thicker, is the ___vitreous___ humor.

6. Excess fluid within the eye is __glaucoma___.

7. Tiny muscles encircling the lens allow it to change __shape____ which in turn allows images to be focused on the __retina__.

8. The __retina_____ is the multilayered surface at the back of the eye which has specialized cells capable of converting light into nerve stimulus.

9. Nerve endings in the retina gather into the ___optic nerve__ which transmits messages to the ___brain_____.

10. The ears have the capability of converting ___sound__ energy into __vibrations_____ which in turn stimulate nerve endings to allow us to hear.

11. Sound moves into the auditory canal and strikes the __tympanic___ membrane which causes the three tiny bones of the middle ear to vibrate. These vibrations stimulate nerve endings within the fluid-filled ___cochlea__ which then send signals by the ___auditory__ nerve to the brain.

12. The semicircular canals within the ear work to maintain __balance___.

13. In our nose there are nerve endings which are capable of taking ___chemical___ stimuli created by contact with tiny molecules of substances floating in the air and converting it to nerve impulses.

14. Stimuli within the nose travels to the brain by the __olfactory__ nerve.

15. The main organ of taste sensation is the __tongue__ . __Papillae_ on the surface of the tongue house taste buds which consist of taste receptor cells.

16. The primary organ of touch is our __skin__. When compared to all other organs in our body, the skin is the __largest__ organ and has the most ___mass___ (weight).

17. The skin consists of ___two____ layers: the _epidermis__ which creates squamous cells that form the surface to the skin and the deeper ___dermis____ which houses blood vessels, nerves, glands and fatty tissue.

18. Touch sensations travel by ___afferent__ neurons to the spinal cord and brain.

19. The skin can be divided into regions of sensation. These regions are known as _dermatomes___.

Name _____Date_____

Lesson 27   Practice Page 2

Use the clues below to solve this crossword puzzle.

Across

2. clear covering of the eye CORNEA

6. largest organ of body enables this sensation TOUCH

10. three tiny bones of middle ear MALEUS INCUS STAPES

14. outer layer of skin with flattened cells EPIDERMIS

15. fluid-filled snail-shaped structure of inner ear with nerve endings inside COCHLEA

16. curved "onion" which focuses light on retina LENS

17. ear drum TYMPANIC MEMBRANE

18. sense of flavor gathered by the tongue TASTE

19. cranial nerve of scent OLFACTORY NERVE

Down

1. light sensitive cell layer in rear of eye RETINA

3. cranial nerve which connects eyes to brain OPTIC NERVE

4. colorful circle in the eye IRIS

5. "job" of the nose SMELL

7. three tiny bones work together to enable this sensation HEARING

8. hole created by the iris to adjust light entry PUPIL

9. our eyes enable this ability SIGHT

11. cranial nerve of hearing AUDITORY NERVE

12. clusters of taste buds on surface of tongue PAPILLAE

13. deeper layer of skin with blood vessels, nerves, glands and fatty tissues DERMIS

Name _____Date_____

Lesson 28   Practice Page 1

Please fill in the blank(s) for each statement below. Use your lesson pages to help you.

1. ___Ecology_____ is the study of the relationships between living things and their environment.

2. _____Environment_ is defined as being everything, both living and nonliving, which are found in the location where a living organism lives.

3. The __biosphere_ is the segment of the earth where life exists.

4. A __biome___ is defined as an area or region of the earth where similar organisms thrive and others do not.

5. The ___polar_____ biome is found at the north and south poles of the earth. The average temperature range is -40 C to -4 C (-40 F to 25 F) with less than 5 inches of precipitation in a year.

6. The temperature range of the __tundra___ is -26 C to 4 C (-28 F to 39 F) and precipitation is less than 10 inches per year. ___Short plants__ and a few mammals are found in this biome.

7. The __coniferous forest__ biome is characterized by evergreen, cone-bearing trees and has a temperature range of this biome is -10 C to 14 C (14 F to 57 F). The average yearly precipitation is between 12 and 30 inches.

8. The _deciduous forest_ biome is characterized by the predominance of trees which are lose their leaves during the cooler seasons of the year. The average temperature of this biome is 6-28 C (42 F to 82 F) and annual precipitation is 30-50 inches.

9. At the same latitude of the deciduous forest biome is the _grasslands_ biome which experiences temperatures slightly cooler than the deciduous forest biomes and receive only 10-30 inches of precipitation in a year. Abundant grasses and few trees are present in this biome.

10. __Desert__ biomes are found in regions which are warmer on average than grasslands yet receive less precipitation. The temperature range in desert biomes are 24-34 C (75 F to 94 F) with precipitation being less than 10 inches per year.

11. The rain forest biome can be divided into the _tropical__ rain forest and the ___temperate__ rain

forest. The tropical rain forest is found all along the equator where the average temperature is 25-27 C (77 F to 81 F). The temperate rain forest is found along the western coast of North America where temperatures range from 10-20 C (50 F to 60 F).

12. The __marine___ biome consists of the earth's oceans and seas.

13. The __freshwater__ biome consists of lakes, rivers and streams. ___Estuaries___ are where rivers of freshwater mix with the salt water of a sea or ocean.

14. The organisms which are capable of capturing energy from the sun and converting it into useable food for itself (and later, others) are known as __autotrophs__. Autotrophs are also called __producers___.

15. Heterotrophs "harvest" energy captured from __autotrophs__. These organisms are also known as _consumers__.

16. The consumers which first eats a producer (plant) is known as a __primary___ consumer. Organisms which eat a primary consumer are called __secondary___ consumers.

17. Consumers which have plants as their sole source of food are known as _herbivores__.

18. Consumers which eat both plants and animals (producers and consumers) are known as ___omnivores____.

19. __Scavengers__ are those organisms which consume dead organisms

20. With each step through an ecosystem away from the producer level, the amount of available energy is __reduced___.

21. __Biogeochemical__ cycles demonstrate how chemical elements move round-and-round through living (bio-) and then non-living portions (geo-) of the environment.

Name _____Date_____

Lesson 28   Practice Page 2

Use the clues below to solve this crossword puzzle.

Down

1. diagram showing how elements move about through the living and non-living segments of the earth BIOGEO-CHEMICAL CYCLE

2. everything, both living and non-living, where an organism lives ENVIRONMENT

3. diagram showing how things eat one another FOOD WEB

4. trees which lose their leaves in the cool seasons of the year DECIDUOUS

5. eater of both plants and animals OMNIVORE

7. study of relationships of living things and their environment ECOLOGY

8. coldest biome of the earth POLAR

9. one who consumes producer CONSUMER

10. another name for autotroph PRODUCER

12. eater of animals CARNIVORE

15. literally means self-feeder AUTOTROPH

16. location where some living things thrive and others do not BIOME

Across

6. eater of road kill SCAVENGER

11. biome which receives most rainfall RAINFOREST

13. biome characterized by abundant grasses and few trees GRASSLAND

14. these are the trees found in the coniferous forest biome CONIFERS

17. you'll only find very short plants and a few mammals living here TUNDRA

18. those who consume autotrophs HETEROTROPH

19. eater of plants HERBIVORE